Culturally Responsive and Sustaining Science Teaching

How can research into culturally responsive and sustaining education (CRSE) inform and transform science teaching and learning? What approaches might teachers use to study CRSE in their classrooms? What are teachers learning from their research that might be transferable to other classrooms and schools?

In this practical resource, teacher researchers from the Culturally Responsive and Sustaining Education Professional Learning Group based in New York City provide insights for educators on how to address complex educational and sociocultural issues in the science classroom. Highlighting wide-ranging and complex problems such as the COVID-19 pandemic and racial injustice and how they affect individual science instruction settings, with a particular focus on urban and high-need school environments, chapters examine and describe what CRSE is and means for science teaching.

Through individual and collaborative research studies, chapters help readers understand various approaches to developing and implementing CRSE strategies in their classrooms and promote students' identification with and affinity for science. Teachers describe the questions driving their investigations, data, and findings, and reflect on their roles as agents of change. Chapters also feature discussion and reflection questions, and include examples of assignments, protocols, and student work that teachers have piloted in their classes.

This book is ideal for pre-service and in-service science teachers and teacher educators across grade levels. It provides support for professional learning activities, as well as undergraduate and graduate teacher education courses. It may be particularly useful in science methods; multicultural education; and diversity, equity, and inclusion courses with a focus on CRSE. This book not only describes one group's approach to CRSE in science education, but also takes the next step to show how CRSE can be applied directly to the science classroom.

Elaine V. Howes teaches curriculum and instruction and educational foundations in the American Museum of Natural History's Master of Arts in Teaching Earth Science Residency Program. She also works with residents and mentors in the program's partner schools.

Jamie Wallace, an educational researcher and evaluator at the American Museum of Natural History, works on the Master of Arts in Teaching Earth Science Residency Program research and evaluation team.

Culturally Responsive and Sustaining Science Teaching

Teacher Research and Investigation
from Today's Classrooms

Edited by Elaine V. Howes and Jamie Wallace

Routledge
Taylor & Francis Group

NEW YORK AND LONDON

Cover photo: © Sean Krepski; Cover concept: Maya Pincus

First published 2024
by Routledge
605 Third Avenue, New York, NY 10158

and by Routledge
4 Park Square, Milton Park, Abingdon, Oxon, OX14 4RN

Routledge is an imprint of the Taylor & Francis Group, an informa business

Funded by American Museum of Natural History.

Library of Congress Cataloging-in-Publication Data
Names: Howes, Elaine V., editor. | Wallace, Jamie, editor.
Title: Culturally responsive and sustaining science teaching : teacher research and investigation from today's classrooms / edited by Elaine V. Howes and Jamie Wallace.
Description: First edition. | New York : Routledge, 2024. | Includes bibliographical references and index.
Identifiers: LCCN 2023050528 (print) | LCCN 2023050529 (ebook) | ISBN 9781032473208 (hbk) | ISBN 9781032503264 (pbk) | ISBN 9781003397977 (ebk)
Subjects: LCSH: Culturally relevant pedagogy--United States. | Culturally sustaining pedagogy--United States. | Science--Study and teaching--Research--United States.
Classification: LCC LC1099.3 .C8475 2024 (print) | LCC LC1099.3 (ebook) | DDC 370.117--dc23/eng/20240130
LC record available at https://lccn.loc.gov/2023050528
LC ebook record available at https://lccn.loc.gov/2023050529

ISBN: 978-1-032-47320-8 (hbk)
ISBN: 978-1-032-50326-4 (pbk)
ISBN: 978-1-003-39797-7 (ebk)

DOI: 10.4324/9781003397977

Typeset in Palatino
by MPS Limited, Dehradun

Access the support material: www.routledge.com/9781032503264

For our New York City students.

About the Cover Image

We have selected this photo for the cover of our book based on several symbolic themes. By showing both trees and the city, it acknowledges the different environments in which our students are learning. The trees represent growth and knowledge. The double exposure resembles a reflection, which in itself is powerful in many ways. As teachers, it is important that we are reflective of ourselves and our practice. As culturally responsive and sustaining educators, we must reflect our students' lived experiences, ideas, and voices in our curricula. This photo was taken by CRSE PLG teacher Sean Krepski.

Contents

Contributors

Arthur W. Funk
USA

Arthur W. Funk works as a peer collaborative coach and teaches Earth science and agricultural sciences at a high school in the Bronx. He is a graduate of the Master of Arts in Teaching Earth Science Residency Program at the American Museum of Natural History. He is currently researching and developing strategies with his colleagues to implement culturally responsive and sustaining practices in science education.

Elaine V. Howes
American Museum of Natural History
USA

Elaine V. Howes teaches curriculum and instruction and educational foundations in the American Museum of Natural History's Master of Arts in Teaching Earth Science Residency Program. She studies how preparation programs support teachers in learning to teach in culturally responsive ways, and how teachers use this learning in their instruction.

Sean Krepski
USA

Sean Krepski is a scientist turned science educator who teaches tenth-grade Earth science at a public high school in Brooklyn, New York. He is a geology graduate of the University of Delaware, and earned his Master of Arts in Teaching Earth Science at the American Museum of Natural History in New York City. In the classroom, Sean focuses on finding ways to connect students' lives to the natural world.

Jamila J. Lyiscott
University of Massachusetts Amherst
USA

Jamila Lyiscott is an associate professor of Social Justice Education at the University of Massachusetts Amherst. There she is the founding co-director of the Center of Racial Justice and Youth Engaged Research. Her community-engaged research examines youth-led activism, race, language, and the capacity of African Diasporic cultures to transgress coloniality.

Contributors

Arthur W. Funk
USA
Arthur W. Funk works as a peer collaborative coach and teaches Earth science and agricultural sciences at a high school in the Bronx. He is a graduate of the Master of Arts in Teaching Earth Science Residency Program at the American Museum of Natural History. He is currently researching and developing strategies with his colleagues to implement culturally responsive and sustaining practices in science education.

Elaine V. Howes
American Museum of Natural History
USA
Elaine V. Howes teaches curriculum and instruction and educational foundations in the American Museum of Natural History's Master of Arts in Teaching Earth Science Residency Program. She studies how preparation programs support teachers in learning to teach in culturally responsive ways, and how teachers use this learning in their instruction.

Sean Krepski
USA
Sean Krepski is a scientist turned science educator who teaches tenth-grade Earth science at a public high school in Brooklyn, New York. He is a geology graduate of the University of Delaware, and earned his Master of Arts in Teaching Earth Science at the American Museum of Natural History in New York City. In the classroom, Sean focuses on finding ways to connect students' lives to the natural world.

Jamila J. Lyiscott
University of Massachusetts Amherst
USA
Jamila Lyiscott is an associate professor of Social Justice Education at the University of Massachusetts Amherst. There she is the founding co-director of the Center of Racial Justice and Youth Engaged Research. Her community-engaged research examines youth-led activism, race, language, and the capacity of African Diasporic cultures to transgress coloniality.

Caity Tully Monahan
USA
Caity Tully Monahan is a graduate of the American Museum of Natural History's Master of Arts in Teaching Earth Science Residency Program. She taught for six years at a high school in Manhattan. Recently, she has collaboratively developed an interdisciplinary curriculum focused on climate change and art. She has also co-developed and co-facilitated teacher professional learning workshops at AMNH on healing-centered pedagogy.

Raghida Nweiran
USA
Raghida Nweiran teaches Earth science and computer science at a charter school in the Bronx. She has spent six years teaching in the classroom, while leading whole grade advisory and coaching the debate team. She is a graduate of the American Museum of Natural History's Master of Arts in Teaching Earth Science, and studies her students' academic behaviors and habits for learning in the classroom.

Maya Pincus
USA
Maya Pincus taught Earth science in public high schools in New York City for six years. She now works as a science communicator for the International Ocean Discovery Program, bringing cutting-edge science to the public all around the world. A geology graduate of the University of Puerto Rico at Mayagüez, and alumna of the American Museum of Natural History's Master of Arts in Teaching Earth Science Residency Program, Maya strives to use her experiences in the field to offer authentic science learning experiences to the public as she has to her students.

Samantha Swift
USA
Samantha Swift is an elementary science cluster teacher in New York City, and taught middle school for six years. After graduating from the American Museum of Natural History's Master of Arts in Teaching Earth Science Residency Program, she spent some time teaching middle school in Los Angeles. Having worked with students across the country and a variety of cultures, she focuses on how to make the classroom relevant to every student she teaches.

Susan Bullock Sylvester
USA
Susan Bullock Sylvester is a graduate of the American Museum of Natural History's Master of Arts in Teaching Earth Science Residency Program. She finished her seventh year of teaching high school Earth science and STEM in a high-need transfer school. Prior to teaching, she had a 28-year career in water resource engineering in Florida. She is interested in how high school students see the work they do in school as relevant to their personal and career goals.

Kin Tsoi
USA
Kin Tsoi teaches Earth science at a large high school in Brooklyn, New York. His student population has changed over time and because of this, his work focuses on teacher–student relationships and classroom culture. His writing is mostly about the building of teacher–student relationships and creating a classroom environment that is conducive to learning and success.

Jamie Wallace
American Museum of Natural History
USA
Jamie Wallace is an educational researcher and evaluator in the education department at the American Museum of Natural History, and works on the Master of Arts in Teaching Earth Science Residency Program research and evaluation team. Her current research focuses on how teachers engage in culturally responsive and sustaining education in science classrooms. She has a background in anthropology and museum ethnography.

Foreword

What does it mean to support an ethnically and racially diverse population of students to feel like science is for them within a schooling system that ignores and seeks to erase their cultural identities as worthless?

When I first encountered the assertion that my culture, and the cultures of people who look and sound and be like me ought to be relevant to a place like school, I was taken aback. As a NYC public school student throughout the 90's and early 2000's the message was clear as day. We took the message in all of its loud silence, and we drank it in deeply. Whoever and whatever we were becomin' through our brown languages, brown cultures, and brown skin should be left at the entryway of the school building, and we could pick it up on our way out, if we so desired. By the time I was a burgeonin' graduate student I had given up on the idea of school as a context where any real justice or equity could be pursued. I stood squarely in my commitment that out-of-school learnin' contexts offered the necessary supplementary and mitigating educational experiences that Black and Brown students urgently needed in the face of a failin' schoolin' system. So when I first came across the work of Gloria Ladson-Billings, who put forth that the complete personhood and dignity of all students should be *relevant* as assets in the classroom, and was later invited into the work of Geneva Gay, who insisted that we unlock the full potential of ethnically diverse students by tailoring who we are as educators to be *responsive* to their respective needs, I felt my first embers of hope. Some decades later, Django Paris and Samy Alim posed a powerful question to us all as they articulated the need for what they called a culturally *sustaining* pedagogy that values linguistic and cultural pluralism. They asked, "What are we seeking to sustain?" *Culturally Responsive and Sustaining Science Teaching* is the book that answers this abiding question with all of the complex nuance that it deserves.

Culturally responsive and sustaining education (CRSE) positions us to stop pretending that our identities can be left at the doorstep when we enter schools and to stop flattening the contours of race and culture in classrooms, but too often this work lingers almost exclusively in the humanities, leaving STEM educators with no concrete tools for bringing it into their classrooms. The teachers in this volume resist the ways that science has become a discipline hyper-focused on old white men as the representational standard, and vulnerably contend with what it means to imbue their classrooms with cultural pluralism that views student resources as assets while maintaining high expectations.

Importantly, this book was written within the realities of an ever-shifting sociopolitical landscape that makes up the backdrop of this volume; a landscape that continuously harms the historically marginalized and underserved communities that necessitate these pedagogical interventions in the first place. That is, the collective of science teachers who came together across five years to understand what culturally responsive and sustaining pedagogies mean for science classrooms were compelled to do this work through the rising visibility of anti-immigration, alt-right, white supremacist groups under the presidency of Donald Trump, through multiple global pandemics that continue to unearth health and economic disparities underscored by racial stratification, and through an ongoing series of extrajudicial public lynchings where the fundamental worth of Black lives is constantly up for political debate.

Most recently, after banning AP African American Studies from Florida schools, on May 15, 2023, Florida Governor Ron DeSantis signed a bill into law that prohibits the state's colleges and universities from funding diversity, equity, and inclusion programs. This is the landscape from which the educators of this volume speak. With both feet on the ground in science classrooms, the teacher researchers here offer the readers tangible insights, pitfalls, tensions, illuminations, and hope for what it means to push the discipline of science education to sustain students' cultural identities as a part of their learning experience. However, please understand that the landscape I am referring to is not *just* a backdrop. It is also the foreground that shows up daily in the work of teaching because it is the immediate reality of students and teachers alike.

For readers wanting to take this up at the level of institutional change, this volume pushes educational institutions to value teacher research as a primary source of wisdom for what is truly needed to transform science teaching in schools. Too often institutional change is guided by research and leadership initiatives helmed by people who are not in the daily minutiae of the very contexts that they seek to transform. Also, by embracing a culturally responsive and sustaining lens, this volume demonstrates a commitment to ensuring that students develop a critical stance to science learning in ways that question the role of institutional power in their educational lives.

For readers wanting to heal the interpersonal relationships between teachers and students, or the broken trust between historically marginalized communities and the schools within them, this volume offers a vulnerable and authentic journey where the authors contend with the relational level of this work. That is, by first naming the realities of their positionalities as mostly white educators teaching ethnically and racially diverse students, the authors invite us into the interpersonal realities of teaching and learning with students and their communities through a CRSE lens. In addition to this, while most teachers lament the isolation of teaching, the authors here have spent eight years in collaboration in ways that showcase the power of collective interpersonal relationships to forge pedagogical development in ways that surpass a required PD here and there.

For those who come to this volume with the understanding that, as Black feminism teaches us, the personal is political, and so take seriously the internal transformation necessary to even contemplate what CRSE needs to be for science classrooms, this volume offers the audacity of introspection and accountability. The kind we need in order to take up the tools offered within each chapter with deep integrity as the authors insist on those things that truly—in this social moment and alongside the agentic students we seek to serve—must unequivocally be sustained in the teaching and learning of science in schools.

In Love & Community,
Jamila J. Lyiscott, aka Dr. J
May 2023

Acknowledgments

We would like to thank our NYC secondary school students for their brilliance, patience, and inspiration. We continue to learn with and from you.

We would also like to acknowledge the administrators and colleagues at the teachers' schools for their support and encouragement as we inquired into culturally responsive and sustaining education in our classrooms.

We would like to thank our colleagues at the American Museum of Natural History (AMNH), with special appreciation to Dr. Maritza Macdonald, Dr. Rosamond Kinzler, Dr. Linda Curtis-Bey, and Dr. Karen Hammerness for ongoing guidance and support over the many years this group has been working together. We extend appreciation to our colleagues in the Master of Arts in Teaching Earth Science Residency Program and in the Educational Research and Evaluation Group for our many long conversations about our work with the CRSE PLG.

We would also like to thank Dr. William E. Rosenthal for challenging and encouraging conversations about teacher research and CRSE, and for his valuable and extensive contributions to references and copy editing.

Special shout-outs to d. alan ward for his remarkable poem, and Toby James for his support and pushing our thinking regarding the cover art. We also extend deep gratitude to all who have helped this book along the way through various contributions and collaborations to further our thinking and learning. Without these, this book would not have come to life. Many thanks to Julia Dolinger and Sophie Ganesh at Routledge for encouraging and supporting us through submission of the manuscript.

Finally, we would like to extend our thanks to the National Science Foundation Robert Noyce Teacher Scholarship Program for financial support for this work through the grant, Preparing and Supporting New Earth Science Teachers through a Museum- and School-Based Teacher Residency Program. This material is based upon work supported by the National Science Foundation under Grant No. DUE-1340006. Any opinions, findings, and conclusions or recommendations expressed in this material are those of the authors and do not necessarily reflect the views of the National Science Foundation.

Acronyms and Abbreviations

ACE	Adverse childhood experience
AMNH	American Museum of Natural History
AMNH ESRP	American Museum of Natural History Earth Science Residency Program
AP	Advanced placement
CER	Claim, Evidence, Reasoning
CRE	Culturally responsive education
CRT	Culturally responsive teaching
CRSE PLG	Culturally Responsive and Sustaining Education Professional Learning Group
CRSE	Culturally responsive and sustaining education
CR-SE	Culturally responsive-sustaining education
CSP	Culturally sustaining pedagogy
DOE	Department of Education
ELL	English language learner
ESRP	Earth Science Residency Program
ICT	Integrated co-teaching
IDS	Interdisciplinary studies
IEP	Individualized Education Program
LA	Los Angeles
MAT	Master of Arts in Teaching
MSA	Muslim Students Association
NGSS	Next Generation Science Standards
NRC	National Research Council
NY	New York
NYC	New York City
NYC DOE	New York City Department of Education
NYS	New York State
NYSED	New York State Education Department
PBAT	Performance-based assessment task
PD	Professional development
PLG	Professional learning group
Pop	Popular

SEL	Social-emotional learning
STEAM	Science, technology, engineering, art, and math
STEM	Science, technology, engineering, and math
TSR	Teacher–student relationship

Introduction

The Culturally Responsive and Sustaining Education Professional Learning Group

Elaine V. Howes and Jamie Wallace

Introduction

If you are reading this book, it is likely that you are wondering: *What does culturally responsive and sustaining education mean for science teaching, and how could it inform my science classroom?* We were intrigued by this question and it became the impetus for our collaboration. This book tells the story of this collaboration through individual and collaborative inquiries based in teachers' questions about CRSE in science teaching.

The goals of this book are to:

- Paint a picture of a multi-year collaboration that may provide valuable insights for beginning and sustaining such partnerships.
- Provide varied images of CRSE in action in science classrooms to meet a growing need for science teaching that engages all students.
- Present a range of examples of questions and methods for teacher research focused on CRSE in science classrooms.
- Suggest a solid foundation for practice, theory, and research of CRSE in science classrooms.

This chapter summarizes how we go about meeting these goals.

DOI: 10.4324/9781003397977-1

The CRSE PLG: Goals and Routines

Our work as a teacher learning community, named the Culturally Responsive and Sustaining Education Professional Learning Group (CRSE PLG), seeks to create engaging science teaching and welcoming classrooms for all of our students. The goals of the CRSE PLG are to explore what culturally responsive and sustaining education (CRSE) could mean for science teaching, develop and implement CRSE strategies in classrooms, and promote students' success in and affinity for science. In this book, we share our individual teacher research studies along with our collaborative journey over the years, including through the early days of the COVID-19 pandemic. By sharing our work, we aspire to contribute to two broad areas of educational research and practice: culturally responsive and sustaining science education, and long-term collaborations as opportunities for teacher learning and growth.

The collaboration that led to this book is crucial to its creation. The science teachers in the CRSE PLG have worked together over a span of eight years, beginning with their preservice teacher residency program, through two years of induction, followed by more than five years in the CRSE PLG. Our time together has allowed for complex conversations about real instruction in real schools as we worked to understand, develop, implement, and study culturally responsive and sustaining approaches. The stories, reflections, and strategies presented here provide starting points for teachers, teacher educators, professional learning facilitators, and researchers interested in science teaching and in collaboration focused in culturally responsive education.

In the early months of the group's work together, it became clear that while there was a lot written about CRSE in literacy and social studies education, there was not the same robust research base for CRSE in science. Although more research relating CRSE to science education has started to appear (see Chapter 1), little focuses on classroom science instruction. And even more notably, *teachers' voices* about grounding their science teaching in CRSE is not prevalent in the literature. We therefore set ourselves the task of applying what we were learning from the broader research base in CRSE to our specific instructional settings.

Teacher Research in Science Classrooms

Intentionally viewing classrooms through our developing CRSE lenses led to specific questions and puzzlements for each teacher and for the group as a whole. These individual and collaborative questions provided the basis for the development of our teacher research studies and chapters. The chapters represent a variety of approaches to inquiry, from in-depth reflection, to designing and noting the outcomes of particular instructional strategies, to recording changes over time in a school's goals and ways of working together. Notably, "reflection" as a research method (Kraft, 2002) comes across in each chapter. We believe that the individuality of each chapter is a strength of this volume.

A good deal of educational research has been written about rather than *with* teachers, and does not necessarily represent teachers' needs. This book presents research conducted by science teachers, with questions, findings, and reflections grounded in their own schools, classrooms, and interactions with students and colleagues. The authors speak from the conviction that teaching is not just a job, but a vocation. Teachers live the complex work of teaching every day—they know better than anyone what they are thinking and why they make the choices they do. The teachers sharing their work in this book say that writing about their inquiries has been a journey in itself, as they summon the courage (not to say the time and intellectual labor!) to share what they are learning with others.

The CRSE PLG has been a forum for teachers to participate in the risky and unfamiliar venture of examining their own teaching, classrooms, and schools. This is something significant about this group that we aim to bring to the fore: First, the group knew each other well and worked together over time. Second, the group could share their teaching experiences (both successful and not) and know they would get sympathetic listeners as well as affectionate challenges to try again, try something different, or merely be prompted to respond to, "What were you thinking when that happened?" This was core to our work together. We discussed literature related to CRSE, and we shared research questions and worked on our inquiries, all in the context of the understanding, forgiving, and challenging space that the group became. This willingness to risk, and to be honest about struggles, is reflected in the teachers' inquiries and writing.

We ask readers to approach this work with the sensitivity and respect that the members of the CRSE PLG extend to one another. We are all learners, teaching is personal work, and writing about it makes us vulnerable. The teachers who have taken this risk hope that readers will do the same in their discussions about their own forays into science teaching based in CRSE, and in studying it with colleagues in their own settings. We hope you will find that this book supports your endeavors to practice, reflect upon, and study culturally responsive and sustaining education in science classrooms.

The Roots of the CRSE PLG

The eight teachers in the CRSE PLG are all graduates of the American Museum of Natural History's Master of Arts in Teaching Earth Science Residency Program (AMNH MAT ESRP) based in New York City (NYC) that prepares science teachers for high-need secondary classrooms. The AMNH ESRP was designed to address the critical shortage of Earth science teachers in New York State. The preservice program begins with a summer museum teaching residency, in which residents learn about teaching and learning in a museum and practice educational strategies with visitors and youth in an informal setting. The program continues with a 10-month residency experience in high-need partner schools, and wraps up with a science research practicum in which residents work with Museum scientists conducting scientific research in the field and in labs. During the 15 months of the program, residents also take graduate-level pedagogy and science courses at the Museum. Following completion of the program, graduates commit to teaching for three years in high-need schools.[1] All of the teacher-authors of this book have completed that commitment and, during the time of their teacher inquiries, were experienced science teachers in NYC high-need schools.

The idea for the group began with the facilitators' suggestion that graduates might want to work together to explore culturally responsive education. As the phrases "culturally relevant pedagogy" and its cousin "culturally responsive education" were in the air, and appeared in professional development activities and at schools, we were interested in exploring what culturally responsive education could look like, specifically in science classrooms. Since this group of graduates continued to teach in NYC, coming together in person was possible and desirable. We met twice a month for a year and a half, reading about CRSE and teacher inquiry, and discussing it in the context of science teaching. Each meeting began with a conversational activity called "Stories from the Field," where teachers told stories of their practice through a CRSE lens (Wallace et al., 2022). As our work together continued, this aspect of the meeting expanded, with teachers sharing situations, strategies, and challenges in their teaching and making connections to the literature. As the group became more confident that what we were learning was valuable for other teachers, we began presenting to preservice and inservice teachers as well as at regional and national conferences.

Over these years, teachers' time has been compensated through grant funding, making participation in this PLG a literal as well as figurative asset for the participants. The commitment to providing this funding indicates that teachers are professionals, their time and expertise is valuable, and when that time is spent on improving their day-to-day work in the classroom, it should be compensated accordingly.

Who We Are and Where We Teach

The group is made up of eight teachers and two facilitators (teacher education researchers), all with connections to the AMNH residency program. During the time of the inquiries in this book, all of the teachers taught secondary science at public or charter high-need schools in NYC that served populations that are majority Black and Latine, made up of students from families who have recently immigrated and those who are long-term citizens. While all of the schools in which we worked during this time are in the same city, they are each distinct, representing different missions and pedagogical approaches, and serving diverse populations of students in multiple boroughs and individual neighborhoods.

The NYC public school system is the largest in the United States. In the 2021–2022 school year, there were 1,058,888 students in the NYC school system, attending 1,859 public schools (NYC DOE, n.d.). According to the U.S. News and World Report (2023), the student population served by the NYC public schools identifies as 41% Hispanic/Latino, 24.7% Black or African American, 16.1% Asian or Asian/Pacific Islander, 14.6% White, 1.9% two or more races, 1.2% American Indian or Alaska Native, and 0.5% Native Hawaiian or other Pacific Islander. Thus, the majority of students in the NYC public school system are students of Color. Additionally, "7.9% of students are English language learners" (U.S. News and World Report, 2023).

Given these data, who we are and how we identify as teachers matters, particularly in our study of CRSE. Therefore, our exploration of our positionality is central to our work together, and our identities provide multiple lenses through which our teaching and our inquiries are communicated. The teachers describe the group as majority white or white presenting, nearly evenly split between cisgender males and females, and unified by the fact that we all work in NYC public schools. We recognize that we do not represent the majority of our students in terms of race, ethnicity, or socioeconomic upbringing. This has indeed been a focus in many of our conversations.

Throughout the chapters, teachers describe the schools where they teach and share insights into their distinct contexts and settings. We see our classrooms and schools as potential sites of change, and our studies illuminate our successes and struggles in grounding our science instruction in culturally responsive and sustaining science teaching. There are parts of our country where teachers and schools are under attack for trying to teach in inclusive and anti-racist ways. Aware of these contemporary debates, we feel privileged to live and teach in an area supported by a public school ethos that welcomes teaching its diverse student populations in the best ways possible. New York City is a remarkably rich context that demands that we develop our thinking and our practice to address rather than suppress the historical inequities inherent in our educational system.

We are a cohesive group with a shared history and a common goal: to use what we are learning to make science education more powerful for the students in our city. We believe that CRSE provides us with a route to do so.

Structure of the Book

Each study in this book is unique, but all are grounded in the exploration of a question related to culturally responsive and sustaining science teaching. In turn, the studies utilize a range of approaches to teacher research. The chapters thus are not prescriptive, and not always generalizable—but designed to inspire discussion and further inquiry and action among readers.

Throughout the chapters, we seek to illuminate the intentionality of each inquiry (Cochran-Smith & Lytle, 1993). As Roberts et al. argue, "The more intentional and systematic the process, the more others perceive these efforts to be research" (2007, p. xvi). Teachers in the CRSE PLG engaged in intentional planning for their inquiries, and used multiple data sources to explore questions inspired by everyday classroom wonderings. Dana and Yendol-Hoppey outlined quality indicators for teacher research to inform self-assessment of inquiries: context of study, wonderings and purpose, teacher-research design (data collection and analysis), teacher-researcher learning, and implications for practice (2009, pp. 173–178). When starting to write up our inquiries, we developed a scaffold of study components that align with these indicators.

This book begins with the chapter *What Do We Mean by Culturally Responsive and Sustaining Science Education? Grounding Our Professional Learning Group in Theory and Research*, which lays out the theory that guides our work. In Chapters 2–9, teachers share investigations stemming from their initial puzzlements and/or goals for their teaching as they were learning about CRSE. Three collaboratively written chapters follow the individual studies. Thus, we present the chapters in four parts, organized by their main purposes or areas of focus: *Exploring Instructional Strategies for CRSE, CRSE and the Science Classroom, School-Based Teacher Collaboration Reflecting CRSE*, and *Collaborative Chapters: Stories and Reflections*.

Framing Chapters

This introductory chapter provides the setting for the work of the CRSE PLG, the nature of the group and its members, and summaries of each chapter.

In Chapter 1, *What Do We Mean by Culturally Responsive and Sustaining Science Education? Grounding Our Professional Learning Group in Theoretical Perspectives and Research*, Wallace and Howes take a deep dive into the theoretical underpinnings of the research and the goals of the group's exploration of CRSE. This chapter is grounded in the literature study we engaged in together during the first years of the group, drawing on the foundational works of Ladson-Billings, Gay, Paris, Alim, and others. The chapter introduces the tenets of CRSE that the group used to develop our shared understanding, and helps to situate this book within the broader literature on teacher professional learning and teacher research.

Part 1: Exploring Instructional Strategies for CRSE

Three chapters focus primarily on teaching strategies inspired by learning about CRSE. In *Dear Dante, Science IS for You: Investigating Student-Driven Discussions in the Science Classroom*, Maya Pincus explores implementing a discussion strategy for current events to address her observation that many of her students enter her science classroom with the notion that science "just isn't for me." In this chapter, she discusses her efforts to facilitate students' development of a "scientist-identity" through learning to speak and think like scientists. She aims to support her students in learning to say, "I am what a scientist looks like!"

Inspired by a definition of the culturally responsive classroom as being centered around students' experiences, Sam Swift writes about incorporating the interests of her middle-school students into a required, scripted science curriculum. In *Pop Culturally Responsive Education: Incorporating Student Interests into a Scripted Curriculum*, Sam describes her efforts to learn more about her students' common interests connected to popular culture, such as video games and anime, and how she used what she learned in conjunction with her curriculum.

In *Helping Students Foster Emotional Connections: Connecting Students' Lives and Communities with the Natural World*, Sean Krepski recalls how he reevaluated his priorities as an educator during the COVID-19 pandemic. In doing so, he changed his teaching approach to focus more on supporting his students in developing emotional connections with science and nature. Sean models making his own connections with the natural world through sharing photographs and samples from his rock collection. Based on this approach, he writes about learning about his students through their sharing of their own stories, photographs, and even the rocks they collected.

Part 2: CRSE and the Science Classroom

In *Behaviors without Behaviorism: Knowing the Students in Room 124*, Raghida Nweiran reflects on common behaviors she sees in her students. Raghida notes that some students take a prompt and run with it, while others wait to be pushed forward by the teacher every step of the way. Through her inquiry, she finds that her students know how they are "supposed to" behave in school, and wonders what she might learn about her students as individuals, to help them more thoroughly take agency in their science learning.

Kin Tsoi reflects on the variety of school cultures he has experienced, having taught at multiple high-need urban schools. In *Humanizing Science Teaching through Building Relationships*, Kin explains his decision to focus on the one thing within his control: his own classroom. He utilizes accounts of classroom interactions and school events to illuminate the importance of conversations with students in establishing a classroom environment, and why they are important to him as a teacher. He includes his students' perspectives on the classroom culture gathered through surveys and reflections on conversations.

In *Building Relationships to Support Relevance: Reflecting on Culturally Responsive Science Teaching,* Susan Bullock Sylvester uses a 3R (Relationships, Relevance, and Reflection) framework to learn about her students' personal goals and hopes for their futures, and considers deeply how she might connect her science teaching to her students' current interests. She shares, along with classroom strategies to support the 3Rs, how she incorporates stories from her own school experiences and earlier career as an engineer to help her identify with her students and their career-oriented goals.

Part 3: School-Based Teacher Collaboration Reflecting CRSE
Caity Tully Monahan writes about the process of integrating trauma-informed practices into her science teaching in *Before CRSE: Trauma-Informed and Healing-Centered Strategies.* She explores the strategies she uses with her students to support building the trust required for CRSE. She also describes a collaboration with school colleagues that entailed creating, implementing, and studying an interdisciplinary curriculum centered on climate change.

Hoping to learn from instructional strategies implemented during the pandemic, Arthur Funk explores collaborative department-wide teacher professional learning in *Crisis Precipitates Change: An Approach to Culturally Responsive and Sustaining Teaching Using a Modified Flipped Classroom.* As a leader of school-based teacher collaboration in his school, Arthur describes his work with his colleagues on how CRSE can inform what they have learned during the pandemic to improve their practice.

Part 4: Collaborative Chapters: Stories and Reflections
The final three chapters in this book were created through the group's collaborative inquiries, as distinct from individual classroom and school-based inquiries: *When the World Tilted Differently: Science Teachers' Pandemic Stories through Culturally Responsive and Sustaining Education* (Chapter 10), *Reflections: Learning In, From, and With the CRSE PLG* (Chapter 11), and *The Ties That Bind* (Chapter 12).

The chapter titled *When the World Tilted Differently: Science Teachers' Pandemic Stories through Culturally Responsive and Sustaining Education* explores our teaching experiences with CRSE in our schools when the COVID-19 pandemic and protests against anti-Blackness hit in 2020. We focus on "Stories from the Field," a conversational routine at our meetings where teachers share observations and insights about CRSE and their science teaching. This chapter describes shifting priorities, relationships with students, and teacher supports during these challenging times.

In *Reflections: Learning In, From, and With the CRSE PLG*, teachers reflect on their experiences of the work of the group over the years. We use this chapter as a space to reflect on and share what we have been learning through our participation in the CRSE PLG. The chapter includes both individual and collaborative reflections, and includes suggestions for science teaching that incorporate aspects of the social, political, and scientific realms through a shared resource to support CRSE in science classrooms.

The last chapter—*The Ties That Bind*—was inspired by the group's noticing that there are common threads that stitch together our inquiries to make an integrated and complex whole. Like the threads connecting the patches on a patchwork quilt, we see these intersecting threads as integral to the purpose of stitching together our documentation, reflection on, and sharing about our collaborative journey into exploring CRSE in our science classrooms. We believe that the CRSE PLG could provide a model for those wishing to embark on teacher research into CRSE. Thus, we also use this space to offer insights and advice for interested educators.

Using This Book for Professional Discussions about CRSE

At the end of each chapter, we provide reflection questions grounded in the respective inquiry described. We hope that these questions will inspire your discussions concerning instructional strategies and research approaches for studying CRSE in your own classrooms, schools, and sociocultural contexts.

With this book, the CRSE PLG opens up our ongoing conversation, welcoming anyone interested in implementing CRSE in their own science classrooms and schools. We hope to hear from you soon.

Note

1 At the time that the CRSE PLG completed the teacher residency program, the teaching commitment was four years in high-need schools in New York State.

References

Cochran-Smith, M., & Lytle, S. L. (Eds.). (1993). *Inside/Outside: Teacher research and knowledge.* Teachers College Press.

Dana, N. F., & Yendol-Hoppey, D. (2009). To collaborate or not to collaborate: That is the question! In *The reflective educators' guide to classroom research: Learning to teach and teaching to learn through practitioner inquiry* (2nd ed.) (pp. 59–71). Corwin Press.

Kraft, N. P. (2002). Teacher research as a way to engage in critical reflection: A case study. *Reflective Practice, 3*(2), 175–189. 10.1080/14623940220142325

New York City Department of Education (NYC DOE). (n.d.). *DOE data at a glance.* Retrieved June 6, 2023, from https://www.schools.nyc.gov/about-us/reports/doe-data-at-a-glance

Roberts, D., Bove, C., & van Zee, E. (Eds.). (2007). *Teacher research: Stories of learning and growing.* NSTA Press.

U.S. News and World Report. (2023). Overview of New York City Public Schools. https://www.usnews.com/education/k12/new-york/districts/new-york-city-public-schools-100001

Wallace, J., Howes, E. V., Funk, A., Krepski, S., Pincus, M., Sylvester, S., Tsoi, K., Tully, C., Sharif, R., & Swift, S. (2022). Stories that teachers tell: Exploring culturally responsive science teaching. *Education Sciences, 12*(6), 401. 10.3390/educsci12060401

1

What Do We Mean by Culturally Responsive and Sustaining Science Education? Grounding Our Professional Learning Group in Theory and Research

Jamie Wallace and Elaine V. Howes

HIGHLIGHTS

- This chapter is designed to address the question of what we mean by culturally responsive and sustaining science education.
- This chapter introduces fundamental tenets of CRSE, drawn from the research, and used by the professional learning group to provide a grounding for the individual and collaborative inquiries in the book.
- With a unique approach to a literature review, the authors take a walk down memory lane reviewing the theoretical perspectives and research that the professional learning group read together in their journey exploring CRSE.
- We hope that people will take away an understanding of the lenses through which we approach CRSE, as well as the foundational scholarship that informed our thinking.

DOI: 10.4324/9781003397977-2

Introduction

This book presents science teachers' inquiries into classroom and school practice grounded in culturally responsive and sustaining education (CRSE). In our approach to teacher research, teachers are "generators of knowledge" (Cochran-Smith & Lytle, 1993b; 2015), and classroom inquiries stem from teachers' own questions, puzzlements, and reflections. Additionally, teacher research aims to provide practical solutions for teachers to adapt for their own classroom use (Stremmel, 2007). Viewing classrooms and schools as sites of change, we draw on Cochran-Smith and Lytle (2015) in our claim that teachers' research can illuminate CRSE in science instruction.

The purpose of this chapter is to shed light on the theoretical perspectives informing our teacher research studies. An important aspect of our work has been reviewing and discussing the literature to develop a shared language and understanding of CRSE. Thus, we take a walk down memory lane, recalling the chronology of where we started in our thinking, how it's changed, and where we are now.

What Do We Mean by Culturally Responsive and Sustaining Education?

Before we broach this question, we need to step back and consider what we mean by culture. "Culture" is a contested term and concept. To support our understanding of "culture," the CRSE PLG read early on an ethnographic history of the term "culture" from the *Encyclopedia of Social and Cultural Anthropology* (Barnard & Spencer, 1996) and an anthropological piece titled "Body ritual among the Nacirema" (Miner, 1956/1991) to gain a deeper understanding about culture through an anthropological lens. Additionally, the concepts of culture and learning each have a complex history (Nieto, 2021). Understanding the role of cultural practices in learning is essential to understanding the foundations of learning (Lee et al., 2020) as it is undergirded in sociocultural perspectives, which emphasize the complexity of learning situated within a specific cultural and social context (Lave & Wenger, 1991; Vygotsky, 1978).

For our purposes, we adopt an interpretation of culture from the New York State Department of Education's Culturally Responsive-Sustaining Education (CR-SE) Framework as "the multiple components of one's identity, including but not limited to: race, economic background, gender, language, sexual orientation, nationality, religion, and ability. Culture far transcends practices such as cuisines, art, music, and celebrations to also include ways of thinking, values, and forms of expression" (NYSED, 2019a, p. 11). We recognize that these components are constantly changing. Taking this a step further, we strongly believe that learning is socially and culturally situated and must be grounded in and center the lived experiences and backgrounds of learners.

As Ladson-Billings and colleagues (2004; 2020) have argued time and again, an important critique of teacher education in the United States is that there is a lack of adequate preparation in terms of experiences and pedagogy concerning culture (e.g., cultural awareness, knowledge, skills, and competencies, practices). Furthermore, much of the research on teacher education adopts an outdated model of teaching that is "culture neutral" (Ladson-Billings, 2000; 2021a), and suggests an emphasis on program structures rather than teaching about culture and specific cultural practices (Ball & Ladson-Billings, 2020; Zeichner & Conklin, 2005). Additionally, examination of one's own culture, identity, and beliefs is often missing from teacher preparation, which can perpetuate dominant white middle-class beliefs and practices, framing them as what is considered to be "the norm" (Ball & Ladson-Billings, 2020).

Why Does CRSE Matter?

The racial and ethnic diversity of students in the United States continues to increase rapidly, while teachers' backgrounds remain largely stable. Thus, there is a pressing need to equip teachers to successfully teach all students. In a recent report, the National Center for Education Statistics (NCES) projected that in Fall 2030 the racial/ethnic backgrounds of primary and secondary students enrolled in public schools throughout the country will see decreases in students identifying as white and Black (projected around 43% and 14%, respectively) with increases in students identifying as Hispanic/Latine, Asian, and of two or more races (projected at 30%, 6%, and 6%, respectively) (NCES, n.d.). In New York, recent findings indicate that while white teachers remained at 80% of the state's teacher workforce from 2011–2017 with Latinx and Black teachers underrepresented, the population of students of Color steadily increased each year to a little below 60% in 2017 (NYSED, 2019b, p. 18).

In our location, New York City (NYC)—home to "the most diverse teacher workforce" in the state—the percent of students of Color was close to 85%, and that of teachers of Color 42% in 2016–2017 (NYSED, 2019b,

p. 20). Thus, the average ratio of teachers of Color to enrolled students of Color was 1:30 (the lowest average ratio in the state) (p. 22). Additionally, as of 2021–2022, nearly 72% of students in the NYC school system were classified as economically disadvantaged (NYC DOE, n.d.). The NYC public school district is also the largest in the nation, with more than 1,800 public schools serving students and families who speak over 182 different languages (Mirakhur et al., 2018).

These data were gathered as part of an initiative to increase the racial diversity and cultural responsiveness of the educator workforce. Research suggests that "cultural synchronicity" between teachers and their students, especially for those from the same racial and ethnic backgrounds, can help bring a deeper understanding of cultural experiences and insights into teaching and learning, resulting in academic benefits (Philip et al., 2017; Villegas & Irvine, 2010). Culturally responsive education developed out of a need to incorporate aspects of students' cultures into teaching and learning, particularly for students experiencing cultural discontinuities because their cultures and backgrounds have been shut out of or omitted from schools. Hence, the teacher needs to play a pivotal role as "cultural accommodator and mediator in promoting student learning" (Nieto, 2021, p. 132).

What's with All the Different Terminologies?

We started the Culturally Responsive and Sustaining Education Professional Learning Group (CRSE PLG) using the terminology of *culturally relevant education.* In our first year as a group, we read foundational works together such as Gloria Ladson-Billings' "But that's just good teaching! The case for culturally relevant pedagogy" (1995a, 2021a). In this piece, Ladson-Billings identified three central criteria of culturally relevant education designed with the success of African-American students in mind. Drawing on multi-year research studies with excellent teachers of African-American students, her theory of culturally relevant pedagogy is based on the criteria that students must experience academic success, develop and/or maintain cultural competence with students' cultures used as a path or mechanism for learning, and develop a critical consciousness to think critically about the world (Ladson-Billings, 1995a, p. 160). In that first year of the CRSE PLG, as we explored literature and shared stories from our classrooms and pondered if something was culturally relevant, inevitably one of us would question, "But isn't that just good teaching?" This concept—"just good teaching"—became a touchstone for our ongoing inquiry into our own instructional choices and strategies, as we attempted to distinguish CRSE approaches from what we currently considered good educational practices.

Ladson-Billings' three criteria of culturally relevant pedagogy provided a clear grounding for this struggle. This piece was, and remains, instrumental in our thinking.

Early on in the history of the group, the "Culturally Responsive Education: A Primer for Policy and Practice" (Johnston et al., 2017) was released from New York University's Metro Center, a group that worked together with the NYC DOE in informing the vision of CRE in NYC schools. We read the primer before our second meeting in September 2018. One of the first of its kind, this primer helped to provide us with an orientation to the research through a policy lens and the framing with which to start exploring our questions: What could this look like in science classrooms? What concrete instructional examples could we put forth? With the release of a 2018 article in the *Gotham Gazette*, we learned that the then NYC mayor, Bill de Blasio, had announced a $23 million investment in culturally responsive education and anti-bias training in schools (Kirkland & Bryan-Gooden, 2018). Understanding that this would be significant for NYC teachers, we pivoted to the terminology of *culturally responsive education* to align our work with the mayor's agenda for NYC schools, as this was the setting for our teaching and learning.

We proceeded to delve into the literature, learning about the multiple iterations and modifications that this pedagogy has undergone since it was introduced by Ladson-Billings and King (1990) and Ladson-Billings (1995a, 1995b, 2014) as culturally relevant pedagogy. A plethora of derivations continue to weave through the research base, such as culturally relevant teaching (Howard, 2001), culturally responsive teaching (Gay, 2010; Hammond, 2014; Souto-Manning, 2018), culturally sustaining pedagogies (Paris & Alim, 2017), culturally responsive education (Johnston et al., 2017), culturally efficacious (Flores et al., 2015; Flores et al., 2018), culturally revitalizing (McCarty & Brayboy, 2021; McCarty & Lee, 2014), culturally thriving (Bang, 2020; Tzou et al., 2021), culturally responsive-sustaining education (NYSED, 2019a), culturally multidimensional (Carter Andrews, 2021), and others. Drawing on earlier work, Ladson-Billings & King (1990) stated, "Culturally relevant teaching is a pedagogy of opposition that recognizes and celebrates African and African-American culture. It is contrasted with an assimilationist approach to teaching that sees fitting students into the existing social and economic order as its primary responsibility" (Ladson-Billings & King, 1990, p. 314).

There was a desire early on in the group's time together to gain clarity as to what culturally responsive education could mean for science teaching. As one teacher said, there were more questions than answers as to what culturally responsive education should look like in science classrooms. In an

Figure 1.1 Word cloud used in an early CRSE PLG meeting in December 2018 to discuss research on culturally relevant and responsive education and culturally relevant pedagogy and the various derivations. Image developed by the authors.

attempt to provide a clearer lens into the field, we created an admittedly oversimplified visual representation (a "word cloud"), using terms that appeared often throughout the scholarship (see Figure 1.1). In terms of gaining vision as to recommendations for science instruction, this was no quick solution, but it did bring to the fore that "this kind of teaching" was focused on a cluster of concepts that socially conscious teachers would endorse (e.g., social justice and equity).

Our shift to the language of *culturally responsive and sustaining education* (CRSE) aligned us with the NYS Education Department's CR-SE Framework (NYSED, 2019a; Wallace et al., 2022). We also found this move appropriate because CRSE incorporates both teaching and learning, and the social-environmental contexts in which they take place. While theoretical perspectives and recommendations concerning culturally responsive teaching emerged in the 1990s (Brown-Jeffy & Cooper, 2011; Gay, 2010; Ladson-Billings, 1995b, 2014; Dodo Seriki & Brown, 2017), more research is needed that provides insights into how to apply CRSE principles into science teaching in concrete ways. Throughout these interpretations of culturally responsive education, several criteria remain central: students' academic success, cultural competence, and development of critical consciousness (Ladson-Billings, 1995a; 1995b), and the integration of students' lived experiences and cultural and language assets and funds of knowledge into instruction (Johnston et al., 2017; López, 2017; Moll et al., 1992). As CRSE conceptualizes teaching as shaped to meet the strengths, needs, interests, and localities of the specific students in the classroom, it is unique to individual students, teachers, and contexts. Nonetheless, central to CRSE is an insistence on "asset-based and student-centered pedagogies of empowerment" (Barron et al., 2022). Also critical to CRSE is the culturally *sustaining* aspect in which connections to one's heritage and culture is maintained with a specific goal of supporting multilingualism and cultural pluralism for students and contribution to social transformation (Paris & Alim, 2017).

Thus, we use the phrase "culturally responsive and sustaining science teaching" to refer to classroom instruction that includes teaching practices to create science classrooms in which all students engage productively in rigorous science learning (Brown, 2019; Mensah, 2019; Windschitl & Calabrese Barton, 2016). We started out with four fundamental tenets of CRE (Howes & Wallace, 2022; Wallace et al., 2022) and have expanded them to encompass our framing of culturally responsive and sustaining science education, in that teaching:

1 values what students bring to the classroom as assets and uses them as resources for teaching and learning (Johnston et al., 2017; López, 2017; NYSED, 2019);
2 draws upon students' cultures to strengthen and sustain their connections to them (Ladson-Billings, 1995a, 2014; Paris & Alim, 2017);
3 holds high expectations for all students' academic learning (Gay, 2010; Ladson-Billings, 1995a); and
4 adopts and supports students in developing a critical stance toward sociopolitical structures and processes (Ladson-Billings, 1995a; Paris & Alim, 2017).

(See Figure 1.2 for a visual representation of our CRSE tenets.)

Our CRSE tenets are not unlike the four principles in the NYSED (2019a) CR-SE Framework outlining 1) welcoming and affirming environment, 2) high expectations and rigorous instruction, 3) inclusive curriculum and assessment, and 4) ongoing professional learning and support. In fact, we see multiple overlaps and intersections across the two frameworks. While organized differently, the main themes across the tenets and principles hold true. These overlaps are not surprising given that the NYSED definition of CR-SE is, "grounded in a cultural view of learning and human development in which multiple expressions of diversity (e.g., race, social class, gender, language, sexual orientation, nationality, religion, ability) are recognized and regarded as assets for teaching and learning" (2019a, p. 10). As Howard (2021) points out, there is a need for more research to better understand how practitioners develop culturally appropriate methods of teaching in classroom settings with culturally diverse populations—a call to which we hope to contribute, and thus further the conversation grounded in classroom practice and drawing on our experiences with our students.

Within the context of our literature review and our group meetings, we discussed the multiple derivations of culturally relevant, responsive, sustaining

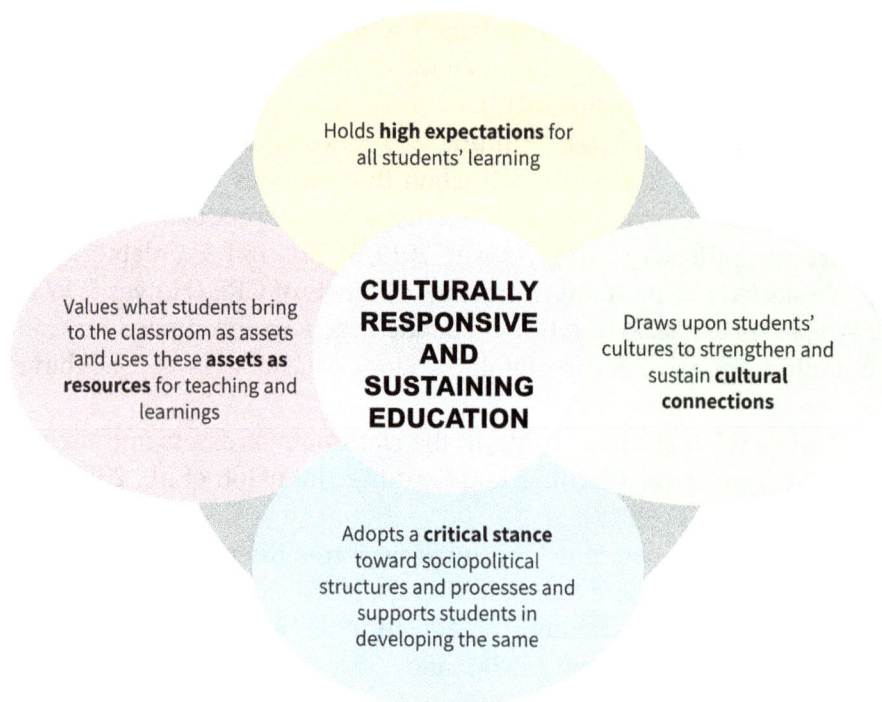

Figure 1.2 CRSE Tenets (Maya Pincus adapted this image from Wallace et al., 2022. © 2022 by the authors. Licensee MDPI, Basel, Switzerland. This article is an open access article distributed under the terms and conditions of the Creative Commons Attribution (CC BY) license.)

pedagogy and education. For instance, while we started off with Ladson-Billings' work on culturally relevant pedagogy in the early 1990s, we continued on to look at Hammond's (2014) *Culturally Responsive Teaching and the Brain*, and examined culturally sustaining education put forth by Django Paris (2012) and expanded into an edited volume by Paris and Alim (2017). The group also dedicated meetings to jigsawing readings where teachers selected texts of interest such as DiAngelo's *White Fragility* (2011), Ladson-Billings' *The Dreamkeepers* (2009/2022), Baines and colleagues' *"We've been doing it your way long enough"* (2018), and National Academies of Sciences, Engineering, and Medicine's (NASEM) second volume on *How People Learn II: Learners, Contexts, and Cultures* (2018) (see Figure 1.3).

Why Culturally Responsive and Sustaining Education in *Science*?

For too long, science education in the United States has been dominated and perpetuated by white eurocentric views contending that science is "neutral," void of culture and bias. The notion that learning science is grounded in

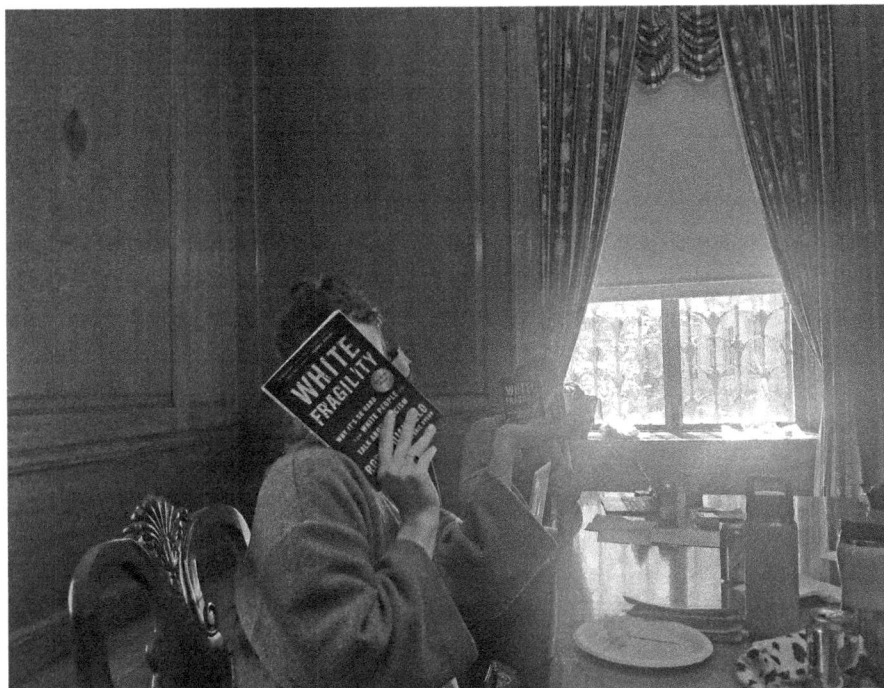

Figure 1.3 Photograph of CRSE PLG members reading and discussing *White Fragility* in May 2019. Photograph by Jamie Wallace.

irrefutable facts and the accumulation of knowledge with clear right and wrong answers is a fallacy that fundamentally contradicts what we know about how people learn and how people *do* science, both of which are steeped in cultural processes. Indeed, both "culture" and "science" have been highly stereotyped and misrepresented concepts that continually need to be challenged (Bang et al., 2013; Bang & Medin, 2010). Recognizing the urgency necessary to reform science teaching and learning, particularly for historically underrepresented students who have been shut out of academic achievement and success, is part of what brings us to culturally responsive and sustaining science education (Norman et al., 2001; Windschitl & Calabrese Barton, 2016; Yerrick & Ridgeway, 2017). Connecting issues of diversity and social justice in science, and consistently questioning "who benefits from the current practice of science," also drives the need for CRSE in science (Laughter & Adams, 2012).

Empirical studies show outcomes and effects on students' learning related to culturally responsive and sustaining practices. Confident in the effectiveness of CRSE supported by the body of empirical evidence, researchers are calling for the institutionalization of CRSE across districts (Parkhouse et al., 2022). In a recent literature review, Sleeter and Zavala (2021) examined six studies from

1988–2007 that focused on culturally responsive teaching, and found that each of the studies had positive outcomes on students' academic achievement and engagement. None of these studies, however, were specific to science.

According to Mensah (2021), culturally responsive approaches in science teaching require "linking students' home experiences to the curriculum, embedding real-world problems in the curriculum, and using examples that connect to students' experiences" (p. 13). It is important that culturally responsive and sustaining teaching is not approached as formulaic or prescriptive, nor can it be considered an add-on or checklist; rather, it is intentionally and thoughtfully planned and enacted for the specific students in the classroom (Ladson-Billings, 2021a; Mensah, 2021; Wallace et al., 2022). Trying to figure out what this can look like in science classrooms was an essential goal of our professional learning group and a main purpose for why we came together.

During the first year of our group, we read chapters from Hammond's (2014) *Culturally Responsive Teaching and the Brain*, a text that was rapidly infiltrating the NYC public schools through the Department of Education. Hammond offers insights into characteristics of dependent and independent learners and argues for a shift toward communities of learners based in her "Ready for Rigor" framework. In thinking about culture, she uses the model of a culture tree to consider surface and deep levels of culture and posits cultural archetypes of collectivism and individualism to help build educators' awareness. Both of these notions became images for discussion in our group. In fact, over the years, we've incorporated an exercise using the culture tree model in developing and leading professional learning experiences for other teachers nationally.

What Does Recent Literature in CRSE in Science Tell Us?

Since our dive into the literature early in our work together, we have focused more on developing our individual and collaborative research studies. Along the way, to support our inquiries, we have kept track of new literature in the field and recognize these contributions to further the field of culturally responsive and sustaining science education.

In traditional Western science education, perhaps more so than other disciplines, Western ways of knowing dominate and are often conveyed and perceived as "neutral" and acultural, divorced from power and bias. While attention to CRSE has been growing nationally, Western science education remains pervasive, even among Native American communities (Higheagle Strong et al., 2023). Scholars have been calling for reform in science education for decades, yet the emphasis and necessity of desettling science education in efforts to make science learning more equitable, accessible, pluralistic, asset-based, relevant, and approachable for all students has been relatively more recent (Bang et al., 2013; Bang et al., 2017; Bettez et al., 2011; Higheagle Strong et al., 2023; Smith et al., 2022; Windschitl & Calabrese Barton, 2016). For instance, Tzou et al. (2021) use the language of *culturally thriving learning and teaching* to express the importance of developing curricula and instructional materials that center equity and social justice in science education.

Emdin et al. (2021) write about the "largely dystopic view of urbanness and the lives of the majority Black youth who attend urban schools" (p. 395) in contrast to the framing of STEM through optimism. They describe the notion of "cultural agnosia," which is "the inability to recognize the value of another's cultural genius as an asset to teaching and learning" and "specifically refer to cultural agnosia as the inability to see hip-hop genius even when the educator can see the genius in self or others" (p. 395).

Bryan Brown (2019) has focused on using students' language assets in STEM classes, emphasizing the importance of centering everyday language in classroom discourse to first tackle the concepts and provide time for students to learn the academic language of science. Highlighting inclusive approaches valuing the diverse languages that students bring into the classroom, Brown introduced instructional practices such as disaggregate instruction and generative formative assessment, as well as the use of technology as a cultural mediator to help "build a bridge between the culture of the teachers and the culture of the students that they serve" (p. 146). Similarly, Lee (2021) calls for an asset-oriented view representing linguistically sustaining pedagogy in science education that maintains high expectations for all learners and supports all students including multilingual learners in rigorous science learning that promotes making sense of phenomena and designing solutions to problems, engages in three-dimensional learning using multiple modalities, and in developing coherent learning progressions over time.

We looked at a couple of pieces that focused specifically on bringing in more of a critical lens in science teaching to support students in developing a critical consciousness. For instance, Boutte et al. (2010) explored ways that a science instructor supplemented science curricula to bring in and center multiple perspectives and counterstories.

Why Teacher Research and Professional Learning?

The first year of the group was dedicated to professional learning and inquiry, reviewing literature related to CRSE, and thinking about what that could look like in science classrooms and instruction. The idea of teacher research was a thread since the first year as well, starting off with chapters by Cochran-Smith and Lytle (1993b) and Dana and Yendol-Hoppey (2009) as readings for our second meeting. We framed these readings with the goal of "getting our discussion about collaborative research going." Both pieces emphasize the importance of shared inquiry among teachers.

Dana and Yendol-Hoppey (2009) highlight the instrumental role of teacher talk when engaging in collaborative research and inquiry in the classroom. This is talk grounded in the daily life of teaching, as well as how teacher inquiry becomes an aspect of each teacher's day. In our group, the conversational routine of Stories from the Field, grounded in teachers' noticings and interactions, has been especially powerful and has expanded over time. In a collaborative paper, we focused on this dimension in particular, analyzing teachers' stories from the first year that the group met (Wallace et al., 2022). In some ways, Chapter 10 in this book can be viewed as a sequel to the 2022 study, focusing particularly on teachers' stories during the time that the COVID-19 pandemic hit NYC.

While the Dana and Yendol-Hoppey chapter was written more recently, they also incorporate quotes and learning from the Cochran-Smith and Lytle chapter, which we returned to several times throughout the years as teacher research projects started to develop. We also discussed human subjects research training, which was followed shortly after by CITI training and certification as a necessary first step when engaging in research involving humans.

Why Are Teacher Professional Learning and Communities of Practice Important?

Research in teacher education indicates that linking theory with practice is possible when professional learning experiences provide opportunities for teachers to reflect upon their teaching in light of the theory they are learning, and, in turn, to use their classroom experience to inform the theory itself. Professional learning groups can provide communities in which teachers can share beliefs, build and construct knowledge, raise questions, and challenge their own assumptions (Cochran-Smith & Lytle, 2015). This collaborative project is rooted in the research indicating that effective professional development for practicing teachers has a common set of features. It is ongoing over a sustained period of time within a community of teacher learners, focused on a particular discipline (e.g., science), and is both grounded in and informing everyday classroom instruction (Bintz et al., 2015; Borko, 2004; Darling-Hammond et al., 2017; Desimone, 2009, 2011; Loucks-Horsley et al., 2009).

Within this community, teachers engage in shared learning around a given area of focus, share resources and experiences, and provide support for one another that is grounded in a process over time (Farnsworth et al., 2016). Such communities of practice provide an important frame for the collaborative sensemaking in which teachers engage in when sharing experiences, insights, and thoughts about their practice and classrooms. Farnsworth and colleagues connect professional communities of practice with a social theory of learning, aiming to "give an account of learning as a socially constituted experience of meaning making" (2016, p. 5). Levine (2010) emphasizes that a teacher professional community focuses on the process of sharing understandings, norms, routines, and trust in relation to teachers' practice with colleagues and students. Teacher communities have a dual concentration on student learning and teacher learning; this inter-relationship informs and shapes teachers' sensemaking in this collaborative setting (Grossman et al., 2001).

In a review of 14 empirical studies investigating the impact of professional learning communities on science teachers, researchers find that participating teachers experience cognitive and affective change, shifts in perspectives, and a collaborative culture focused on improving practice (Dogan et al., 2016). Research shows that design features for effective teacher professional learning incorporates models that are closely tied to practice, content specific, require active participation, are collaborative in nature, sustained over time, and include reflection on classroom practice (Darling-Hammond et al., 2017).

When the CRSE PLG was launched, teachers were beginning their third year of teaching and had known each other well for more than three years as members of the same cohort in their teacher residency program and throughout two years of new teacher induction. Therefore, they had shared experiences, and were able to discuss ideas and feelings openly as well as challenge each other in supportive and respectful ways. Additionally, because they all taught similar content in the same city, they could design, implement, and critique teaching strategies collaboratively.

Why Is Teacher Research Important?

We ground our vision of teacher research in the Cochran-Smith and Lytle assertion "that teacher research makes accessible some of the expertise of teachers and provides both university and school communities with unique perspectives on teaching and learning" (1993a, p. 23). This broad statement denies the claim that the only valuable knowledge is that constructed by "objective" researchers from outside of the classroom (Cochran-Smith & Lytle, 1993b). While teacher research has historically been marginalized

(Lytle & Cochran-Smith, 1992), teachers are "generators of knowledge" and their research is grounded in everyday puzzlements (Cochran-Smith & Lytle, 1993b; 2015) and wonderings about phenomena observed in the classroom (van Zee, 2006). CRSE PLG voices and experiences provide the foundation for our explorations of and inquiries into CRSE in science classrooms, drawing on our own reflections and questions, and designed to help further teaching practice or provide solutions to practical everyday problems (Stremmel, 2007).

Teacher research is similar to teaching in that it accentuates "local knowledge" as an invaluable asset, and is informed by relationships and school cultures and communities, thus blurring boundaries between research and practice (Cochran-Smith & Lytle, 2015). It also encourages critical thinking about one's own teaching practices and students' learning, contributes to the research enterprise, and informs reform-based instruction (Kincheloe, 2003; Roberts et al., 2007; van Zee, 2006). Teacher research can work toward social change and disrupt narratives of inequities, emphasizing the importance of the teacher as an agent in sharing their research to inform broader audiences. As scholars note, teacher research is "based on a deep and profound sense of accountability for students' learning and life changes" (Cochran-Smith & Lytle, 2015, p. 58). Acknowledging and advocating that teachers are indeed among, if not *the*, most important factor in educational change (Cochran-Smith & Lytle, 2015), teacher research should be uplifted and considered as generative contributions to the field. Taking it a step further, teacher research can also be viewed as products of change, as the process of conducting the research itself furthers thinking and practice.

Considering teacher research particularly in science education, Roth (2007) posed important questions to the field. She asked, "What kind of knowledge is being produced by science teacher research? In what ways might this knowledge be of interest beyond the individual teacher researcher and her/his immediate collaborative group or school?" (p. 1,228). In this book, we hope that we have provided a response to these questions.

To explore examples of teacher research in our group, we looked to work in science education. For instance, Mary Bell's (2007) chapter investigating a puzzlement of why some of her students in her second-grade classroom were not talking proved to be a particularly useful sample and something to which the teachers in our group could relate. In her chapter, Bell clearly documented her learning from her various data collection activities including reviewing student records, oral inquiry, tracking progress using an action plan, journaling, and tallying how many times one particular student talked in class. Later, we read Souto-Manning

Figure 1.4 Some of the books that informed our thinking about culturally responsive and sustaining education and teacher research. Photograph by Jamie Wallace.

et al.'s (2018) *No More Culturally Irrelevant Teaching*, a collaboration between teachers and teacher educators focused on examining CRE in elementary classrooms. This book certainly informed our thinking and in some ways served as an exemplar for the book you are reading now. (Figure 1.4 portrays some of the books on CRSE and teacher research that informed our group thinking.)

What Informed Our Thinking about Teaching and Practitioner Research during the Pandemic?

Several pieces became especially influential in our thinking during the pandemic. As a group, during school closures, we delved back into the literature as a bit of an escape from what we were experiencing on a daily basis. The teachers in the group developed a list for our group reading, and we read and discussed works such as Baldwin's (1963/2008) "A Talk to Teachers," Coates's (2015) *Between the World and Me*, Delpit's "The Silenced Dialogue" (1988) and *Teaching When the World is on Fire* (2019), and Noah's (2016) *Born a Crime*. When reading Coates' book, we listened to sections in the audio recording to hear his voice as he read aloud his story. The power and emotion that he expressed in his letter to his son helped us in homing in on centering students' voices and the immense strength in uplifting our students' voices. Similarly, we listened to portions of Trevor Noah's book aloud together during our meetings, to hear the inflections and nuances he conveyed as he shared his story.

As a group, during the pandemic, we also spent time talking about healing-centered pedagogy and the shift from trauma-informed care (Ginwright, 2018). As one of our teachers led a virtual professional development series for other teachers focused on healing-centered engagement with the return to school in 2020 (VanBiene & Tully, 2020), we found ourselves returning to the topic often and considering ways it informed our practice and classrooms with a priority on social-emotional learning, noting connections to CRSE. Especially at the start of the pandemic, focusing on how to build empathy and promoting well-being (moving away from a common deficit view of trauma), and the embedded connections to culture and identity resonated with us and our work both in teaching as well as in our teacher research. As Ginwright wrote, "Healing centered engagement uses culture as a way to ground young people in a solid sense of meaning, self-perception and purpose. This process highlights the intersectional nature of identity and high-lights the ways in which culture offers a shared experience, community and sense of belonging" (2018). (See Chapter 8 for more on healing-centered pedagogy.) Indeed, Howard (2021) identified a pedagogy of care as integral to CRSE, naming as one of five key principles: "an authentic and culturally informed notion of *care* for students, wherein their academic, social, emotional, psychological, and cultural well-being is adhered to" (p. 140).

As the pandemic progressed and the science developed, providing us with vaccines and deeper understanding of how the virus spreads, we were able to share our individual and group inquiries in their preliminary stages at virtual and then in-person conferences including the annual meetings of the Ethnography Forum at the University of Pennsylvania (pre-pandemic), National Association of Research in Science Teaching (NARST), NSF Noyce Summit, and American Educational Research Association (AERA). We were honored to have Dr. Jamila Lyiscott as our discussant during our structured poster session at the 2022 AERA annual meeting (Wallace & Howes, 2022). Having read *Black Appetite, White Food* together in our group, Dr. J's feedback and structured conversation helped push our thinking forward. For us, an integral piece of her book is the chapter titled, "If You Think You're Giving Students of

Color a Voice, Get Over Yourself!" (Lyiscott, 2019). The importance of affirming students' expertise in its endless facets that come into play in the classroom and making it explicit came through to us as a critical element of CRSE. While this is work that we had already been doing and is "just good teaching," highlighting the importance of it and viewing it as one component of CRSE resonated with us.

As schools opened and in-person teaching resumed, we switched from books to articles, including Love's (2020) "Teachers, We Cannot Go Back to the Way Things Were" and Hammond's (2021) "Liberatory Education: Integrating the Science of Learning and Culturally Responsive Practice." Both of these articles speak to the ongoing and persistent systemic inequalities that our students face on a daily basis and deficit ideologies undergirding our school systems, and emphasize the urgency of new and counter-narratives. They also both highlight the importance of teachers as agents of change. These articles were instrumental in our discussions about our practice, and even played a role in some of the teachers' shifts in their own inquiries. In thinking about our teacher research during this time, Cochran-Smith and Lytle's (2015) words resound even louder in considering the role that practitioner research can play in contributing to and producing local, on the ground knowledge, both to inform the broader public as well as to advocate and promote wider social change and educational equity.

Providing a Theoretical Grounding for the Chapters That Follow

We hope that this chapter offers context for the forthcoming inquiries, illuminating and grounding our work in decades of research. Throughout each of these bodies of research (teacher professional learning, communities of practice, teacher research), the common denominator remains the power and positioning of teachers as agents of change. Coupled with the focus on CRSE and science, it was inevitable that a longstanding experience like this would culminate in some type of lasting product that could be shared more broadly. Then, being hit by an international pandemic provided even more impetus, motivation, and a needed boost to get this book out and into the hands of teachers, aspiring teachers, teacher educators, administrators, and teacher researchers. In doing so, we embrace a vision of good science teaching that centers equitable and just practices including pedagogy that is culturally responsive and sustaining.

> **Reflection and Discussion Questions**
>
> 1 What powerful books or articles have you read about CRSE that you would recommend to others starting their journey? Why?
> 2 Which scholars have really pushed your thinking about equity pedagogy? In what ways?
> 3 *Teachers:*
>   In what ways have you engaged in teacher research or practitioner research? Why was it important? What did you learn?
>   **OR**
>   If you have not yet done this, what are you interested in finding out? What questions from your everyday practice are you interested in exploring in a deeper and more systematic way?
>
>   *Teacher Educators:*
>   In what ways have you worked with practicing teachers on teacher research? What were some of the important lessons for you in this work?
>   **OR**
>   If you have not done this, where might you start?

References

Baines, J., Tisdale, C., & Long, S. (2018). *"We've been doing it your way long enough": Choosing the culturally relevant classroom.* Teachers College Press.

Baldwin, J. (2008). A talk to teachers. In M. Cochran-Smith, S. Feiman-Nemser, D. J. McIntyre, & K. E. Demers (Eds.), *Handbook of research on teacher education: Enduring questions in changing contexts* (3rd ed.) (pp. 202–207). Routledge. 10.4324/9780203938690 (Reprinted from "A Talk to Teachers," 1963, December 21, *Saturday Review*).

Ball, A. F., & Ladson-Billings, G. (2020). Educating teachers for the 21st Century: Culture, reflection, and learning. In N. S. Nasir, C. D. Lee, R. Pea, & M. McKinney de Royston (Eds.), *Handbook of the cultural foundations of learning* (pp. 387–403). Routledge. 10.4324/9780203774977-27

Bang, M. (2020). Learning on the move toward just, sustainable, and culturally thriving futures. *Cognition and Instruction, 38*(3), 434–444. 10.1080/07370008.2020.1777999

Bang, M., Brown, B., Calabrese Barton, A., Rosebery, A. S., & Warren, B. (2017). Toward more equitable learning in science. In C. V. Schwarz, C. Passmore, & B. J. Reiser (Eds.), *Helping students make sense of the world using next generation science and engineering practices* (pp. 33–58). NSTA Press.

Bang, M., & Medin, D. (2010). Cultural processes in science education: Supporting the navigation of multiple epistemologies. *Science Education, 94*(6), 1008–1026. 10.1002/sce.20392

Bang, M., Warren, B., Rosebery, A. S., & Medin, D. (2013). Desettling expectations in science education. *Human Development, 55*(5–6), 302–318. 10.1159/000345322

Barnard, A., & Spencer, J. (1996). *Encyclopedia of social and cultural anthropology*. Routledge.

Barron, H. A., Brown, J. C., & Cotner, S. (2022). The culturally responsive science teaching practices of undergraduate biology teaching assistants. *Journal of Research in Science Teaching*, *58*(9), 1320–1358. 10.1002/21722

Bell, M. P. (2007). When students don't talk: Searching for reasons. In D. Roberts, C. Bove, & E. van Zee (Eds.), *Teacher research: Stories of learning and growing* (pp. 44–57). NSTA Press.

Bettez, S. C., Aguilar-Valdez, J. R., Carlone, H. B., & Cooper, J. E. (2011). On *negotiating White science*: a call for cultural relevance and critical reflexivity. *Cultural Studies of Science Education*, *6*, 941–950. 10.1007/s11422-011-9355-1

Bintz, J., Hvidsten, C., Kowalski, S. M., Numehahl, P., Roth, K. J., Stennett, B., Taylor, J. A., Wickler, N. I. Z., & Wilson, C. (2015, April 11–14). *Testing the consensus model of effective PD: Analysis of practice and the PD research terrain* [Conference presentation abstract]. Eighty-ninth Annual NARST International Conference, Chicago, IL, United States. https://narst.org/sites/default/files/2019-07/2015_Abstracts.pdf

Borko, H. (2004). Professional development and teacher learning: Mapping the terrain. *Educational Researcher*, *33*(8), 3–15. 10.3102/0013189X033008003

Boutte, G., Kelly-Jackson, C., & Johnson, G. L. (2010). Culturally relevant teaching in science classrooms: Addressing academic achievement, cultural competence, and critical consciousness. *International Journal of Multicultural Education*, *12*(2). 10.18251/ijme.v12i2.343

Brown, B. A. (2019). *Science in the city: Culturally relevant STEM education*. Harvard Education Press.

Brown-Jeffy, S., & Cooper, J. E. (2011). Toward a conceptual framework of culturally relevant pedagogy: An overview of the conceptual and theoretical literature. *Teacher Education Quarterly*, *38*(1), 65–84.

Carter Andrews, D. J. (2021, July). Preparing teachers to be culturally multidimensional: Designing and implementing teacher preparation programs for pedagogical relevance, responsiveness, and sustenance. *The Educational Forum*, *85*(4), 416–428. 10.1080/00131725.2021.1957638

Coates, T.-N. (2015). *Between the world and me*. New York: Spiegel & Grau.

Cochran-Smith, M., & Lytle, S. L. (1993a). Learning from teacher research: A working typology. In *Inside/Outside: Teacher research and knowledge* (pp. 23–40). Teachers College Press.

Cochran-Smith, M., & Lytle, S. L. (1993b). *Inside/Outside: Teacher research and knowledge*. Teachers College Press.

Cochran-Smith, M., & Lytle, S. L. (2015). *Inquiry as stance: Practitioner research for the next generation*. Teachers College Press.

Dana, N. F., & Yendol-Hoppey, D. (2009). To collaborate or not to collaborate: That is the question! In *The reflective educators' guide to classroom research: Learning to teach and teaching to learn through practitioner inquiry* (2nd ed.) (pp. 59–71). Corwin Press.

Darling-Hammond, L., Hyler, M. E., & Gardner, M. (with assistance from Espinoza, D.). (2017). *Effective teacher professional development*. Learning Policy Institute. https://bibliotecadigital.mineduc.cl/bitstream/handle/20.500.12365/17357/46%20Effective_Teacher_Professional_Development_REPORT.pdf?sequence=1

Delpit, L. (1988). The silenced dialogue: Power and pedagogy in educating other people's children. *Harvard Educational Review, 58*(3), 280–299. 10.17763/haer.58.3.c43481778r528qw4

Delpit, L. (Ed.). (2019). *Teaching when the world is on fire.* The New Press.

Desimone, L. M. (2009). Improving impact studies of teachers' professional development: Toward better conceptualizations and measures. *Educational Researcher, 38*(3), 181199.

Desimone, L. M. (2011). A primer on effective professional development. *Phi Delta Kappan, 92*(6), 68–71. 10.1177/003172171109200616

DiAngelo, R. (2011). White fragility. *International Journal of Critical Pedagogy, 3*(3), 54–70. https://libjournal.uncg.edu/ijcp/article/view/249/116

Dodo Seriki, V., & Brown, C. T. (2017). A dream deferred: A retrospective view of culturally relevant pedagogy. *Teachers College Record, 119*(1), 1–8. 10.1177/016146811711900101

Dogan, S., Pringle, R., & Mesa, J. (2016). The impacts of professional learning communities on science teachers' knowledge, practice and student learning: A review. *Professional Development in Education, 42*(4), 569–588. 10.1080/19415257.2015.1065899

Emdin, C., Adjapong, E., & Levy, I. P. (2021) On science genius and cultural agnosia: Reality pedagogy and/as hip-hop rooted cultural teaching in STEM education. *The Educational Forum, 85*(4), 391–405. 10.1080/00131725.2021.1957636

Farnsworth, V., Kleanthous, I., & Wenger-Trayner, E. (2016). Communities of practice as a social theory of learning: A conversation with Etienne Wenger. *British Journal of Educational Studies, 64*(2), 139–160. 10.1080/000071005.2015.1133799

Flores, B. B., Claeys, L., & Gist, C. D. (2018). *Crafting culturally efficacious teacher preparation and pedagogies.* Lexington Books.

Flores, B. B., Claeys, L., Gist, C. D., Clark, E. R., & Villarreal, A. (2015). Culturally efficacious mathematics and science teacher preparation for working with English learners. *Teacher Education Quarterly, 42*(4), 3–31.

Gay, G. (2010). *Culturally responsive teaching: Theory, research, and practice* (2nd ed.). Teachers College Press.

Ginwright, S. (2018, May 31). *The future of healing: Shifting from informed care to healing centered engagement.* Medium. https://ginwright.medium.com/the-future-of-healing-shifting-from-trauma-informed-care-to-healing-centered-engagement-634f557ce69c

Grossman, P., Wineburg, S., & Woolworth, S. (2001). Toward a theory of teacher community. *Teachers College Record, 103*(6), 942–1012. 10.1111/0161-4681.00140

Hammond, Z. (2014). *Culturally responsive teaching and the brain: Promoting authentic engagement and rigor among culturally and linguistically diverse students.* Corwin Press.

Hammond, Z. (2021). Liberatory education: Integrating the science of learning and culturally responsive practice. *American Educator, 45*(2), 4–11. www.aft.com/summer2021/hammond

Higheagle Strong, Z., Charlo, L. J., Watson, F., Price, P. G., & Christen, K. (2023). Weaving together Indigenous and Western knowledge in science education: Reflections and recommendations. *Journal of Indigenous Research, 10*, Article 10. https://digitalcommons.usu.edu/kicjir/vol10/iss2022/10

Howard, T. C. (2001). Telling their side of the story: African-American students' perceptions of culturally relevant teaching. *The Urban Review*, *33*(2), 131–149. 10.1023/A:1010393224120

Howard, T. C. (2021). Culturally responsive pedagogy. In J. A. Banks (Ed.), *Transforming multicultural education policy & practice: Expanding educational opportunity* (pp. 137–163). Teachers College Press.

Howes, E., & Wallace, J. (2022, August 9). Exploring culturally responsive science teaching through turbulence and challenge: Starting a multi-year research study during the pandemic. *AAAS Advancing Research & Innovation in the STEM Education of Preservice Teachers in High-Need School Districts (ARISE)*. https://aaas-arise.org/2022/08/09/exploring-culturally-responsive-science-teaching-through-turbulence-and-challenge-starting-a-multi-year-research-study-during-the-pandemic/

Johnston, E. M., D'Andrea Montalbano, P., & Kirkland, D. E. (2017). *Culturally responsive education: A primer for policy and practice*. Metropolitan Center for Research on Equity and the Transformation of Schools, New York University. https://steinhardt.nyu.edu/metrocenter/culturally-responsive-education-primer-policy-and-practice

Kincheloe, J. L. (2003). *Teachers as researchers: Qualitative inquiry as a path to empowerment* (2nd ed.). Routledge. 10.4324/9780203497319

Kirkland, D. & Bryan-Gooden, J. (2018, May 3). The mayor's much-needed investment in culturally responsive education. *Gotham Gazette*. https://www.gothamgazette.com/130-opinion/7653-the-mayor-s-much-needed-investment-in-culturally-responsive-education

Ladson-Billings, G. (1995a). But that's just good teaching! The case for culturally relevant pedagogy. *Theory Into Practice*, *34*(3), 159–165. 10.1080/00405849509543675

Ladson-Billings, G. (1995b). Toward a theory of culturally relevant pedagogy. *American Educational Research Journal*, *32*(3), 465–491. 10.3102/00028312032003465

Ladson-Billings, G. (2000). Fighting for our lives: Preparing teachers to teach African American students. *Journal of Teacher Education*, *51*(3), 206–214. 10.1177/0022487100051003008

Ladson-Billings, G. (2004). It's not the culture of poverty, it's the poverty of culture: The problem with teacher education. *Anthropology & Education Quarterly*, *37*(2), 104–109. 10.1525/aeq.2006.37.2.104

Ladson-Billings, G. (2014). Culturally relevant pedagogy 2.0: aka the remix. *Harvard Educational Review*, *84*(1), 74–84. 10.17763/haer.84.1.p2rj131485484751

Ladson-Billings, G. (2021a). *Culturally relevant pedagogy: Asking a different question*. Teachers College Press.

Ladson-Billings, G. (2022). *The dreamkeepers: Successful teachers of African American children* (3rd ed.). John Wiley & Sons.

Ladson-Billings, G., & King, J. (1990). *Cultural identity of African-Americans: Implications for achievement*. Midcontinental Regional Education Laboratory.

Laughter, J. C., & Adams, A. D. (2012). Culturally relevant science teaching in middle school. *Urban Education*, *47*(6), 1106–1134. 10.1525/aeq.2006.37.2.104

Lave, J., & Wenger, E. (1991). *Situated learning: Legitimate peripheral participation.* Cambridge University Press.

Lee, C. D., Nasir, N. S., Pea, R., & McKinney de Royston, M. (2020). Introduction: Reconceptualizing learning: A critical task for knowledge-building and teaching. In N. S. Nasir, C. D. Lee, R. Pea, & M. McKinney de Royston (Eds.), *Handbook of the cultural foundations of learning* (pp. xvii–xxxv). Routledge. 10.4324/9780203774977

Lee, O. (2021). Asset-oriented framing of science and language learning with multilingual learners. *Journal of Research in Science Teaching, 58*(7), 1073–1079. 10.1002/tea.21694

Levine, T. H. (2010). Tools for the study and design of collaborative teacher learning: The affordances of different conceptions of teacher community and activity theory. *Teacher Education Quarterly, 37*(1), 109–130.

López, F. A. (2017). Altering the trajectory of the self-fulfilling prophecy: Asset-based pedagogy and classroom dynamics. *Journal of Teacher Education, 68*(2), 193–212.

Loucks-Horsley, S., Stiles, K. E., Mundry, S., Love, N., & Hewson, P. W. (2009). *Designing professional development for teachers of science and mathematics.* Corwin Press.

Love, B. L. (2020). Teachers, we cannot go back to the way things were. *Education Week, 29.* www.edweek.org/leadership/opinion-teachers-we-cannot-go-back-to-the-way-things-were/2020/04

Lyiscott, J. (2019). *Black appetite. White food.: Issues of race, voice, and justice within and beyond the classroom.* Routledge.

Lytle, S. L., & Cochran-Smith, M. (1992). Teacher research as a way of knowing. *Harvard Educational Review 62*(4), 447–474.

McCarty, T., & Lee, T. (2014). Critical culturally sustaining/revitalizing pedagogy and Indigenous education sovereignty. *Harvard Educational Review, 84*(1), 101–124.

McCarty, T. L., & Brayboy, B. M. J. (2021). Culturally responsive, sustaining, and revitalizing pedagogies: Perspectives from Native American education. *The Educational Forum, 85*(4), 429–443.

Mensah, F. M. (2019). *Teaching culturally and ethnically diverse learners in the science classroom [White paper].* McGraw Hill Education. https://www.mheducation.com/unitas/school/explore/research/teaching-culturally-and-ethnically-diverse-learners.pdf

Mensah, F. M. (2021). Culturally relevant and culturally responsive: Two theories of practice for science teaching. *Science and Children, 58*(4), 10–13. https://www.nsta.org/science-and-children/science-and-children-marchapril-2021/culturally-relevant-and-culturally?fbclid=IwAR0pF75q_M_0JIbQqrpVgQIQiP-vFDdHqizUBU5iHdwh-uMItymwpQjao_U

Miner, H. (1991). Body ritual among the Nacirema. In A. Podolefsky & P. J. Brown (Eds.), *Applying cultural anthropology: An introductory reader* (pp. 20–23). Mayfield (Reprinted from "Body ritual among the Nacirema," 1956, *American Anthropologist, 58*[3], 503–507. 10.1525/aa.1956.58.3.02a00080)

Mirakhur, Z., Sludden, J., Solanti, J., & McGuinness, S. (2018). *NYC public schools: What does it mean to be the nation's largest district?* The Research Alliance for New York City Schools.

https://steinhardt.nyu.edu/research-alliance/research/spotlight-nyc-schools/nyc-public-schools-what-does-it-mean-be-nations

Moll, L. C., Amanti, C., Neff, D., & Gonzalez, N. (1992). Funds of knowledge for teaching: Using a qualitative approach to connect homes and classrooms. *Theory Into Practice, 31*(2), 132–141. 10.1080/00405849209543534

National Academies of Sciences, Engineering, and Medicine. (2018). *How people learn II: Learners, contexts, and cultures.* National Academies Press. https://nap.nationalacademies.org/catalog/24783/how-people-learn-ii-learners-contexts-and-cultures

National Center for Education Statistics. (n.d.). *Racial/Ethnic enrollment in public schools.* U.S. Department of Education, Institute of Education Sciences, National Center for Education Statistics. Retrieved January 13, 2022, from https://nces.ed.gov/programs/coe/indicator/cge

New York City Department of Education. (n.d.). *DOE data at a glance.* https://www.schools.nyc.gov/about-us/reports/doe-data-at-a-glance

New York State Education Department (NYSED). (2019a). *Culturally responsive-sustaining education framework.* http://www.nysed.gov/crs/framework

New York State Education Department (NYSED). (2019b). *Educator diversity report.* http://www.nysed.gov/common/nysed/files/programs/educator-quality/educator-diversity-report-december-2019.pdf

Nieto, S. (2021). Culture and learning. In J. A. Banks (Ed.), *Transforming multicultural education policy & practice: Expanding educational opportunity* (pp. 111–136). Teachers College Press.

Noah, T. (2016). *Born a crime: Stories from a South African childhood.* Doubleday Canada.

Norman, O., Ault, C. R., Jr., Bentz, B., & Meskimen, L. (2001). The black-white "achievement gap" as a perennial challenge of urban science education: A sociocultural and historical overview with implications for research and practice. *Journal of Research in Science Teaching, 38*(10), 1101–1114. 10.1002/tea.10004

Paris, D. (2012). Culturally sustaining pedagogy: A needed change in stance, terminology, and practice. *Educational Researcher, 41*(3), 93–97. 10.3102/0013189X12441244

Paris, D., & Alim, H. S. (Eds.). (2017). *Culturally sustaining pedagogies: Teaching and learning for justice in a changing world.* Teachers College Press.

Parkhouse, H., Bennett, E., Pandey, T., Lee, K., & Johnson Wilson, J. (2022). Culturally relevant education as a professional responsibility. *Educational Researcher, 51*(7), 474–480. 10.3102/0013189X221092390

Philip, T. M., Rocha, J., & Olivares-Pasillas, M. C. (2017). Supporting teachers of color as they negotiate classroom pedagogies of race: A study of a teacher's struggle with "friendly-fire" racism. *Teacher Education Quarterly, 44*(1), 59–79.

Roberts, D., Bove, C., & van Zee, E. (Eds.). (2007). *Teacher research: Stories of learning and growing.* NSTA Press.

Roth, K. J. (2007). Science teachers as researchers. In S. K. Abel & N. G. Lederman (Eds.), *Handbook of research on science education* (pp. 1,205–1,259). Erlbaum.

Sleeter, C. E., & Zavala, M. (2021). What the research says about ethnic studies. In J. A. Banks (Ed.), *Transforming multicultural education policy & practice: Expanding educational opportunity* (pp. 209–238). Teachers College Press.

Smith, T., Avraamidou, L., & Adams, J. D. (2022). Culturally relevant/responsive and sustaining pedagogies in science education: theoretical perspectives and curriculum implications. *Cultural Studies of Science Education, 17*(3), 637–660. 10.1007/s11422-021-10082-4

Souto-Manning, M., Llerena, C. L., Martell, J., Maguire, A. S., & Arce-Boardman, A. (2018). *No more culturally irrelevant teaching.* Heinemann.

Stremmel, A. J. (2007). The value of teacher research: Nurturing professional and personal growth through inquiry. *Voices of Practitioners, 2*(3), 1–9. https://citeseerx.ist.psu.edu/document?repid=rep1&type=pdf&doi=d65c79cc4fad7379697e813e06cd017664a2a540

Tzou, C., Bang, M., & Bricker, L. (2021). Commentary: Designing science instructional materials that contribute to more just, equitable, and culturally thriving learning and teaching in science education. *Journal of Science Teacher Education, 32*(7), 858–864. 10.1080/1046560X.2021.1964786

VanBiene, N., & Tully, C. (2020, August 6). *Healing centered engagement & a start to school [Workshop].* American Museum of Natural History.

van Zee, E. (2006). Teacher research: Exploring student thinking and learning. *Science Educator, 15*(1), 29–36. https://citeseerx.ist.psu.edu/document?repid=rep1&type=pdf&doi=641d39fdc8069f73e21039c0e7dd6145e92b7b1e

Villegas, A. M., & Irvine, J. J. (2010). Diversifying the teaching force: An examination of major arguments. *Urban Review, 42*(3), 175–192. 10.1007/s11256-010-0150-1

Vygotsky, L. S. (1978). *Mind in society: The development of higher psychological processes.* Harvard University Press.

Wallace, J., & Howes, E. (2022). *Teacher research into culturally responsive-sustaining science teaching: Not going back to how things were before.* [Chairs and presenters] Structured poster session presented at the American Educational Research Association (AERA) Annual Meeting. April 2022. San Diego, CA.

Wallace, J., Howes, E. V., Funk, A., Krepski, S., Pincus, M., Sylvester, S., Tsoi, K., Tully, C., Sharif, R., & Swift, S. (2022). Stories that teachers tell: Exploring culturally responsive science teaching. *Education Sciences, 12*(6), 401. 10.3390/educsci12060401

Windschitl, M., & Calabrese Barton, A. (2016). Rigor and equity by design: Locating a set of core teaching practices for the science education community. In D. H. Gitomer & C. A. Bell (Eds.), *Handbook of research on teaching* (5th ed., pp. 1099–1158). American Educational Research Association. 10.3102/978-0-935302-48-6

Yerrick, R., & Ridgeway, M. (2017). Culturally responsive pedagogy, science literacy, and urban underrepresented science students. In M. Milton (Ed.), *Inclusive principles and practices in literacy education* (pp. 87–103). Emerald Publishing Limited. 10.1108/S1479-363620170000011007

Zeichner, K. M., & Conklin, H. G. (2005). Teacher education programs. In M. Cochran-Smith, & K. M. Zeichner (Eds.), *Studying teacher education: The report of the AERA panel on research and teacher education*, 645–736. Lawrence Erlbaum Associates Publishers.

Part

1

Exploring Instructional Strategies for CRSE

2

Dear Dante, Science IS for You

Investigating Student-Driven Discussions in the Science Classroom

Maya Pincus

HIGHLIGHTS

- In this chapter, Maya focuses on the CRSE tenet of developing a critical consciousness through the question: *In what ways can student-led current events discussions help foster a "scientist identity" in my classroom?*
- In a letter written to a fictitious student (Dante), Maya notes that Dante entered her classroom thinking science "isn't for him." Through the letter, she strives to convince him that he *is* a scientist already.
- Maya describes an experiment with student-led discussions as a way to help her students develop a scientist identity in the classroom.
- We hope you take away from this chapter a concrete way to support student-led discussions that students find engaging and supports them in thinking of themselves as scientists.

DOI: 10.4324/9781003397977-4

Introduction and Context

Teachers, how many times have you heard a student express that they hate science, or that they "just can't do it"? All too often, high school students come into my Earth science classroom with a fixed idea that science is something separate from them, and that it is cold and unrelatable. For me, this has come up in all seven years of my teaching career, in the three schools I've taught in Brooklyn and the Bronx.

In this chapter, I am writing to my students who feel that they have been wronged by science. This is what I wish I could explain every time a student claims, "Science just isn't for me!" The letter below is addressed specifically to one student, yet Dante represents the many students who share his feelings—both those who have expressed them to me outright, and those who might be too shy to do so. In my letter, I explore the question: *In what ways can student-led current events discussions help foster a "scientist identity" in my classroom?*

Most recently, I've been teaching Earth science in a high school in Brooklyn, serving predominantly Hispanic, Latinx, and Black students. About 25% of the students at my school qualify for special education services, and almost 30% of my students are classified as English language learners (NYC DOE, n.d.). Some students immigrated to the United States as recently as a few weeks before they joined my class. This is the school context in which my inquiry takes place.

Preconceptions of Science

My students bring a wide range of experiences and backgrounds to the classroom. I have spent many years trying to track down the source of the feelings that my students often express that science "just isn't for me." As is usually the case, the range of stories is as broad as the number of students I teach. But in my persistence to get to the heart of the issue, I have identified a few themes in my students' past experiences with science that have led them to this point.

For some, science has been portrayed as something dry and boring, a process kept hermetically sealed inside a distant laboratory. For others, the idea of who can be a scientist is chained to pictures of Albert Einstein and Walter White. Additionally, a disconnect between the natural elements in the mandated Earth science curriculum and students' lived experiences growing up in a city can make it challenging for teachers to make the content accessible. In a lot of ways, our educational system has wronged young people when it comes to science: for instance by focusing on rote memorization of facts, and emphasizing the importance of aged white men as role models. It is no wonder that many feel alienated from this field.

Before I became a teacher, I was following the typical academic track. After receiving my bachelor's degree, I immediately accepted an offer to a master's program at the University of Puerto Rico–Mayagüez. One important part of academia, especially in science, is frequent attendance at conferences to disseminate research findings and network with people involved in similar lines of investigation. At one conference, I had the opportunity to truly pause and look around. Almost every single person I saw was white, and many were men, middle-aged or older. Up until this point, I had mostly been able to float through my life unaware of my whiteness. I was living the definition of white privilege. But observing the demographics of this science conference, especially considering perspectives of my Latinx peers, I finally saw that science needed a face lift.

I can't help but wonder to what extent my identity affects how my students feel in science class. My students like me because I'm kind and I treat them with respect, but I don't feel that they find me particularly relatable. I am a white woman and young enough that students don't often ask me if I have kids, but too old to understand why they care so much about social media. I was raised in the suburbs by an almost perfectly "traditional" family. I went to college without paying a cent, thanks to a modest merit scholarship and a fund set up by my grandfather before I was born. Given that my hair is curly, I tan easily, and I'm fluent in Spanish, I can occasionally "pass" as Latina; but the moment I open my mouth any pretext disappears. I often wonder if I'm making the problem worse when I show up to my students as another white scientist.

The conversations that I've had with students, those frequent reminders that many of them believe that science is not for them, are what led me to culturally responsive and sustaining education (CRSE). It is not just about students feeling that they are represented in my class and in the science content, but about empowering them to believe that the content I teach is specifically for them. It is not only important that they learn the content but that they can use what they learn to become agents of change to make our world better for everyone.

Something I've had to explore throughout my dive into CRSE is the tension between the students and the system they are forced into. It is undeniable that the very structure of our education system is rooted in inequity, and my students have been disadvantaged by this their whole lives. However, I feel very strongly that this should not be an excuse to not hold them accountable to high expectations for learning. Part of what makes education culturally responsive and sustaining is the ways in which it holds students to high expectations. This goes hand in hand with science, in a way that I endeavor to help my students see that even when they are unsuccessful on their first (or fifth) attempt, they should be proud to learn from their mistakes and try again.

In this chapter, you'll read about how I attempted to leverage the tenets of CRSE to encourage my students to not hate science. More than that, I wanted them to have opportunities to build a relationship with science, and feel like science was something they have the power to do. I knew it would be hard for students to feel that they *looked* like scientists, so I explored other ways to support the development of a scientist identity. I thought if students could get comfortable *talking* about science, maybe that could be a stepping stone to engage them. I wondered if I could transform students' relationship with science by having them choose current events articles about topics that interested them, and then facilitate conversations about the articles with their peers. I spent a year compiling and analyzing classroom artifacts, student surveys, discussion participation data, and my own reflections to answer my research question on fostering a "scientist identity" in my classroom through student-led discussions on current events.

My use of CRSE is not formulaic. Rather, my approach is to provide students with enough opportunities to think, speak, and act like scientists that by the end of the year they leave my classroom with the perspective "I AM a scientist!" This comes from how I decorate my classroom, the videos I choose to show, the articles I have students read, and even the way I design my students' learning experiences. My goal is to help students feel comfortable enough in my classroom that they are willing to lean in when something does not go as planned, and know that working through a problem, making a mistake, and trying again is what it means to be a scientist.

Join me on my journey of learning what it means to be a culturally responsive and sustaining science teacher. I hope that you benefit from hearing about what did and did not work with my students, and that you can find at least one strategy that will empower your students to want to embrace their own identities as young scientists. Thank you for your dedication to our children.

To my dearest Dante, You walked into my classroom on the first day of school, and before I even had time to greet you, you said "Ms., no disrespect, but science just isn't my thing." You smiled warmly and took your seat, and I knew that despite my dedication to care for you this year, I had already lost you. At least in the beginning.

I'll start by letting you know that I see you and I hear you. You are not the first person to walk into my classroom feeling this way, and I'm sure you won't be the last. Soon I'll tell you about the things some of my previous students have told me, about why they thought science wasn't for them either. But for now, I want to beg you to not give up so soon! Even though you don't realize it yet, science *is* for you. You *are* a scientist!

Remember that time you were trying to get more likes on your Instagram posts? You thought maybe if you included your dog in your next photo, you'd get more likes. So you took the picture, posted it, and waited for the response. What did you learn? Did you include your dog in the picture after that because you found that it worked? You may not believe me, but you are a scientist. You were wondering, "How do I increase my Instagram likes?" That was your question. You thought pictures of your dog might help. Hypothesis. You posted the picture and waited to see what happened. Boom! Experiment.

And what about that time you needed to figure out how to make sure your alarm got you up in time to get to school? You told me the problem was that you kept turning the alarm off instead of hitting snooze. But, you fixed it by wrapping your sister's scrunchie around the part of the phone where it said "Stop," so you couldn't hit it by accident anymore. Engineers get *paid* to come up with these kinds of solutions!

Still not convinced? I get it. I understand why you don't see yourself as a scientist, especially since I have known many other students who feel the same way.

Developing a Scientist Identity

Has anyone ever explicitly told you that you can be a scientist? That being a scientist is a cool, prestigious thing? Remember not too long ago when all you wanted was to be an astronaut when you grew up? What happened to change that?

A couple of years ago, I asked students in my Earth science class if they knew any scientists. I kept the question broad by choosing to not specify what I meant by "scientist," to get an idea of my students' interpretations of that word. Of the 48 tenth graders who were in class the day I gave the survey, 19 said that I was the only scientist they knew. (While I'm flattered to hear that students remembered the stories about my life as a researcher before I transitioned to teaching, the few photos they've seen don't really do justice to my time in the field.) Three students had family members they considered scientists: two nurses and a psychologist. Two students also mentioned the famous scientists Isaac Newton and Bill Nye. The rest of my students said that they did not know any scientists, and I have yet to meet a student who personally knows anyone involved in the hard sciences (chemistry, geology, physics … you get it). It's like there's some sort of invisible barrier between SCIENCE and REAL PEOPLE.

There's a reason I gave my students this survey. Research has shown time and again that people of Color are underrepresented in STEM fields (Dutt, 2020; Goldberg, 2019; NCSES, 2021; Rathbun et al., 2018). This is a national problem, but I was curious to what extent it was reflected among the students in my class. I was hoping that if my students already had connections with scientists, I could leverage those relationships to increase engagement in my class. Unfortunately, the survey results were even more bleak than the national level. In the USA, only 20% of Earth and space science professionals are people of Color (NCSES, 2021). But out of the people represented by my students' networks, it's 0%! Not a single one of my students reported having a personal connection to a professional Earth scientist. If we opened my classroom data up to all fields of science rather than just geology, it's still only 4%.

I can imagine that it might be hard to see yourself as a scientist if you've never met one, and possibly have never seen a scientist that looks like you. If you've done a Google search for "famous scientists" any time before 2023, you were instantly presented with a monotone grid: nine faces, all white, all but two male (see Figure 2.1a). The search results have been slightly improved since then, now featuring three women and one Black scientist (Figure 2.1b).

You might have been told by teachers in your life that "You can be anything you want once you put your mind to it!" But at the same time, images you've seen and messages you've heard may have conveyed the opposite. According to New York State Education Department statistics, there are about 60% students of Color in the state, while about 80% of your teachers have been white (NYSED, 2019a). And I wouldn't be surprised if at least one of your previous science teachers had tried to inspire you by hallowing their classroom walls with posters of esteemed scientists. Were any of those scientists in the posters people of Color? Or did they represent the ghosts of our nation's racist past? Were you inspired by these photos to become a scientist? Or did you want to distance yourself from their white gaze (Ilmi, 2011; Yancy, 2013) glaring down at you from beyond the grave?

An activity I did in class last year showed me that these experiences can certainly make an impression on young learners. Early in the school year, I asked my students to illustrate what a scientist looks like. The results were almost unanimous: In just my first-period Earth science class, 17 out of 19 students drew or selected images of a white person or a small group of white people in a

(a)

(b)

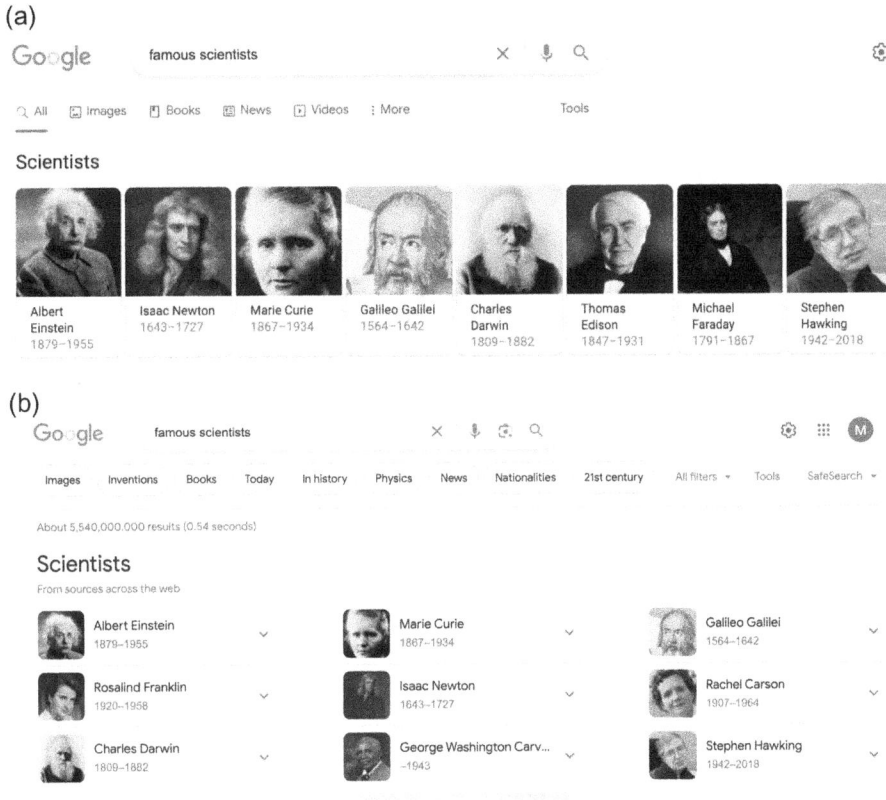

Figure 2.1a and 2.1b A screenshot from a simple Google search for "famous scientists" confirming the bias that the typical scientist is an old white man. Downloaded from Google December, 2021. Screenshot by the author.

laboratory. (See Figure 2.2 for an example of one of my student's projects that describes that a scientist is "a white man who looks kind of creepy.") The results in my other classes were similar. When I was a kid, it was Doc Brown and Dexter's Laboratory. For you, it's Sheldon from the *Big Bang Theory* and still, almost every Google search result. I can infer from these data that many of my students see science and scientists as *other*—not something often practiced by people who look like them. And I'm not the first science teacher to do this experiment. Since 1947—NINETEEN FORTY-SEVEN!!!—researchers have been asking students to draw what a scientist looks like (Mead & Métraux, 1953). Guess what students drew back then? How much do you think it has changed in 70 years? The answer is disappointing (Finson, 2002). It's devastating to be an educator in 2023 and to see almost no change.

Even when I actively try to choose content that represents you and your culture, and elevates marginalized voices, it's a challenge. I sometimes spend hours searching for stories and videos that demonstrate that some of

When I think of scientist I think of a white man who looks kind of creepy. I also think of mixing chemicals and studying space. I think of nature and robots/technology.

Figure 2.2 An example of a student project depicting what they believed a scientist looks like. In this image, the student wrote that she thinks of a scientist as "a white man who looks kind of creepy." Screenshot by the author.

the coolest scientists are Black and Brown. When will the day come that I can search for stories about scientists and get diverse results, without specifically seeking them out? When will I no longer have to search for "famous Black scientists," and "famous Latinx scientists," because scientists of Color will automatically be included in the results?

That's where we come in. We are the ones who can make this change. This might sound insane to you now, Dante, but by the time you make it to the end of this letter, I'm hoping you'll be convinced that you *are* a scientist, and that science *is* for you. And when you embrace that, you will be a scientist role model who can inspire other students for years to come.

From a Teacher's Perspective

Give me a moment to talk to you about this from a teacher's perspective. Educators have known for *years* that students perform best in school when their teachers get to know their students well, and teach in a way that makes connections to what students know and like (Brown, 2007; Irvine & Armento, 2001; Ladson-Billings, 2001; Villegas & Lucas, 2002). But that just sounds like good teaching, right (Ladson-Billings, 1995)? I hope that when you see yourself and your culture represented in my classroom, the content will be way more meaningful to you. For example, which water cycle lesson sounds better? Memorizing vocabulary from diagrams and reiterating it word for word back on a test, *or* investigating Flint, Michigan, as a case study, researching not just how political negligence led to a health crisis, but also about diverse activists who fought to protect their community? You would choose the second option, right? Of course you'd rather learn something that has obvious meaning and real-world applications.

Years of research into effective teaching tells us that if you, Dante, can connect to what I'm teaching, you are more likely to feel agency in your learning, and therefore be more likely to want to engage and excel (NASEM, 2018). This concept has been known as culturally relevant pedagogy (Baines et al., 2018; Ladson-Billings, 1995), culturally responsive education (Hammond, 2014), culturally sustaining pedagogy (Paris, 2012), and more recently culturally responsive and sustaining education, or CR-SE (NYSED, 2019b). Whatever name we settle on, it is something I am trying to learn how to do well.

This approach started in social studies and English language arts (ELA) classrooms. We know now that we need to teach multiple sides of history, and leave traditions of colonialism behind. Every single country in the world has its wealth of scholars, artists, and innovators. Why do our history books focus only on white Americans of European ancestry? Why do these history books ignore that our country was founded on the backs of enslaved people? America would be nothing of what it is today if it weren't for the Black and Brown people who were forced to work with no way out (Hannah-Jones, 2019; Reynolds & Kendi, 2020). And to bring it back to STEM, let's not forget that our modern study of mathematics is based on advances that came from Arabic intellectuals in northern Africa and the Middle East (Falagas et al., 2006).

Thinking critically about multiple perspectives in history probably isn't news to you. However, as far as traditional American education goes, the idea to include this information in school was revolutionary. Once teachers started teaching that history was MADE by people of Color, the students demonstrated a more complex understanding (Epstein et al., 2011). It's similar in ELA classrooms. Why are groups of young people of Color forced to read about a little blond girl who broke into a bear's house (i.e., *Goldilocks and the Three Bears*), or generationally rich white men who throw exorbitant parties in the Hamptons (i.e. *The Great Gatsby*), when they could be reading about a Dominican-American boy struggling to find his place in life as he grows up in *The Brief and Wondrous Life of Oscar Wao* (Díaz, 2008), or a Black girl learning how to love herself despite racism even from within her family in *Genesis Begins Again* (Williams, 2020)?

It's inspiring to hear about these educational shifts that move us toward celebrating achievements across all cultures. With CRSE, students are not only exposed to role models they can relate to, but students and teachers alike can learn through these examples that diversity is power. This diversity-is-power thing is clearly evident in countless global organizations. For one, the International Tsunami Information Center relies on the collaboration of experts from places like Chile, Indonesia, Portugal, Japan, and Senegal to develop the most effective global strategies to mitigate tsunami risk (Pararas-Carayannis, 1988). Scientists from over 190 countries around the world came together to write the Paris Agreement, the first step in a global effort to prevent devastating climate change (United Nations, n.d.).

These examples show us that diversity of all types is integral to scientific progress. People from each nation bring different perspectives that, when combined, allow the team to consider problems from multiple angles and develop multi-dimensional solutions. Differences should be celebrated—demanded—in both the science classroom and the professional world. It's refreshing to see that the movement is beginning to take hold (Medin & Lee, 2012). Now, we need scientists like you, because you bring a unique perspective to finding creative solutions to our society's problems.

A Culturally Responsive and Sustaining Science Classroom

How am I supposed to be culturally responsive and sustaining in my science classroom? Isn't my job to teach you how to memorize information, follow strict procedures in a lab, collect data, and draw conclusions that are unbiased by your personal experiences? Where is there room for your—or anyone's—personality and culture when science is all about using facts to explain phenomena? This stereotype is probably why the research into what makes a culturally responsive and sustaining science classroom was so limited when I started learning about it. And I wonder … How have you been affected by the stereotype that science is so impersonal? Is that part of what drove you away? Even when a novelist who doesn't look like you writes about something you've never experienced, at least you can imagine what it might be like to live their character's life.

Science is done by people, for people, in an effort to answer questions about the natural world. And the way we answer questions is through science. Take the Moon for example. Every year or so you probably see a certain "supermoon" start to trend. Last year it was the Super Strawberry Moon, this year it's the Worm Supermoon. Did you know that these names come from Native Americans, who gave the full moon a different name each month to connect it to their own experiences as the seasons passed (National Geographic Staff, 2019)? For example, the Anishinaabe, Cherokee, and Mahican tribes named the Strawberry Moon for when the fruit comes into season (American Indian Alaska Native Tourism Association, n.d.), and the Naudowessie tribe named the Worm Moon for when the emergence of beetle larvae signals the beginning of spring (Almanac, n.d.).

And what about your star sign? Dante, I know you're a Sagittarius, but do you know why? Even before there was written language, there were scientists! We have cave paintings that show people were trying to connect the movement of stars in the night sky to observable phenomena on Earth as early as 25,000 years ago (Campion, 2008). See? Science is so much more about humans and our experiences than most people think. And the best part? Science is for everyone, and should be done by everyone.

Becoming a Scientist in the Classroom

Going back to what you told me on the first day of school, that you "just can't do science," I've been spending a lot of time thinking about how we can work together to rewrite that narrative in our class this year. I hope that over the course of this school year you will not just have the opportunity to do amazing science, but to discover and internalize that you are a scientist.

For now, let me ask you: What would it take for you to be excited to approach novel problems, because it's an opportunity to work towards a solution? As your teacher, I am dedicated to finding ways that you can empower yourself through science. Remember how you thought a scientist is someone who sits alone in a lab mixing together different chemicals? In reality, scientists are so much more.

- Scientists Are Collaborators
  It's a great story when you hear about the one guy who made a ground-breaking discovery all on his own, but in reality scientists work in teams that require frequent, productive conversation.

- Scientists Are Communicators
  To share what they've learned with the world, scientists speak at conferences and on the news, publish articles, and can be some of the best public speakers around. Scientists even use social media platforms like Instagram and TikTok to broadcast information and network with colleagues.
- Scientists Are Problem Solvers
  There are so many things wrong with the world, from hunger to pandemics to droughts. When faced with these issues, scientists brainstorm the kinds of solutions that can change the world. And almost none of these amazing results were achieved on the first try. What truly makes a scientist is the ability to try, fail, and then try again.

But enough with my motivational speech. What can I actually *do* to help you feel more comfortable taking up a scientist identity? As I mentioned, I'm grounding my teaching in something called CRSE to help you feel that science *is* for you. I'm trying to teach in ways that value what you and my other students bring to the table, draw on your cultures to help you feel connected to the content, hold high expectations for everyone, and help you think critically about socio-political structures and processes (Howes & Wallace, 2022; Wallace et al., 2022). With these objectives, I am hoping that you see yourself as a science scholar, and take agency in your role as a scientist.

The first step in this process was figuring out the source of the problem: Was I the only teacher with some students saying that they "just can't do science," or were other science teachers observing similar things in their classrooms? I regularly meet with a team of teachers from other schools around our city, and all of them have reported that they too have students who shy away from science. We started meeting to discuss literature and figure out how to improve our teaching practices. It quickly became clear that before we could do anything, we had to answer a fundamental question: *What does a culturally responsive and sustaining science classroom look like?*

Bringing Theory to Life

We started with this idea of "cultural relevance." To us, that means making connections to references that you would get. If I try to get you thinking about rocks by describing the tombstones in *Buffy the Vampire Slayer*, that's going to fall pretty flat (though maybe it would work for your parents). But if I start my minerals unit by mentioning the diamonds on Jay-Z's watch, and show pictures of Drake's marble bathtub to introduce metamorphic rocks, it is more likely that you'll know what I'm talking about. I can also make my class more relevant when I teach about latitude and longitude. For instance, I can have you learn to plot locations on the map based on cities where you and your classmates have family. It helps to ground new information in the things you already know, but it's still not enough.

This brings us to the idea of culturally responsive education. If we go back to what you've observed about science and scientists in the past, it's evident that this field is whitewashed by history. On one hand, that means that white Europeans are being glorified as the heroes of science. But on the other hand, that means that people of Color have been *excluded* (Fields, 1998). Not just hidden from history, but shut out completely. It's almost as if we are supposed to believe that people of Color *can't* or *shouldn't* do science. If I want you to feel like you belong in my class, I need to respond to this crime of history. That goes way beyond the content of my mandated Earth science curriculum. This isn't something you'll see on your standardized test at the end of the year, but is an important part of the critical stance we both need to take to rewrite the narrative.

With this in mind, my teacher group began to consider the idea of student identity in science class. We knew that it would be an uphill battle trying to convince our students that they look like scientists, especially given the persisting racial inequities in the field. Instead, we speculated we might help our students begin to see themselves as scientists if they could *think* and *talk* like scientists. And so, we agreed that culturally responsive education in the science classroom could be characterized by student voice.

We all had different ideas about what student voice should look like. In this context, when I use the term "student voice," I am not simply talking about how many times you open your mouth, or the noise levels in our classrooms. I am referring to your participation in academic conversations in which you demonstrate critical thinking using evidence and analysis. When it comes down to it, that's all science is! Science is the process of using observations and reasoning to answer a question or solve a problem. If we can help our students develop those abilities, and make it clear that engaging in those practices is what scientists do (NRC, 2012), then you will have no fuel to support the argument that you "can't do science."

Thus, our hypothesis was born: If we can help our students develop their "scientist voices," then they will be able to identify as scientists. If our students identify as scientists, they will be more likely to pursue higher education and a STEM career. All we had left to do was test our hypothesis. (Even though I'm a teacher now, I'm still a scientist, too!) We also had different ideas about how to foster student voice in our classrooms.

Cultivating student voice in my science classroom

For me, there were four important considerations to keep in mind regarding how to foster student voice in my classroom that helped me figure out what strategy to use.

1 **It had to be something that gave my students the opportunity to make their own choices.** Obviously, an assignment is going to be more engaging if it's something students choose. More than that, choosing can put students in charge of their learning, requiring them to take agency in the classroom and ownership of learning and success (Baroutsis et al., 2016; Morrison, 2008).

2 **It had to be something that positioned my students to practice being responsible.** As a teacher, it is not just my job to teach facts about Earth science. One of my most important roles is helping students learn how to be successful once they are out of school.

3 **It had to be grounded in real science.** Many of my students think of science as something that "other" people do in "other" places. If we could learn about diverse examples of science in the real world, it would be easier for students to make connections between science and their own lives.

4 **It had to provide the opportunity to have academic conversations about science.** The point of this was to help my students feel comfortable having their voices heard in a scientific setting, so that they could develop the confidence necessary to see themselves as scientists.

Bringing Current Events Discussions into My Classroom

To meet these criteria, I came up with a weekly current events assignment. Each week, one or two students were responsible for leading the activity. My plan was to keep it fair by using a random name generator to determine the order of presenters. However, I know that no one wants to be the first person to do a brand-new thing. I incentivized going first by allowing anyone who volunteered to present early on to work with a partner of their choice. After the volunteers were scheduled, I randomly assigned the remaining students individually.

The recurring assignment took place every Friday. First, the presenter had to find a news article that talked about something related to Earth science. I know that this might not be intuitive without practice, so before we started, I took the class to the computer lab and provided a list of websites that had reputable articles about Earth and space science. In addition to *The New York Times* and *CNN*, the list included news websites specifically for students, and in multiple languages for my emergent bilingual and multilingual learners. I modeled using the search function on the websites to find an article that interested me, and how to read for the main points of the article. Students then had time to practice on their own. Theoretically, everyone who came to class that day should have then known how to access an article of their choice online.

This aspect of the assignment, the student choice part, really excited me because I was giving my students the opportunity to seek out articles that interested them. I anticipated that students would be genuinely engaged in the content, since they picked it. Even more, this allowed me to better understand what engaged my students, and I planned to use that information to inform each unit throughout the year. For example, several students in a row chose articles related to global warming, so I knew that I could draw out my climate unit to focus on the intricacies of climate change, including examples of young people in the forefront of climate activism. Another reason that this element of student choice was important to me was that I believed it would be an opportunity for students to find something they could connect to.

The weekly leader was expected to share their article with me by Wednesday, so that I could share it with the class and students would have time to read it before the discussion. The rest of the students were expected to come to class on Friday having read the article. The leader's role was to come prepared with several open-ended questions to facilitate a whole-class discussion. I thought this would be a chance for my students to practice public speaking, so that they could develop a confident voice in a professional academic setting.

Class on Friday was dedicated to the current events discussions. I did not expect the conversations to last the full period—I told my students that I expected to spend about 20–25 minutes on that activity—so the rest of class would be used for quizzes or housekeeping. Before I go into what actually happened, I will describe my idealized version.

In an Ideal World ...

If everything were to go perfectly, the weekly leader would send me their article by noon on Wednesday, so I could post it on our class's online learning management system before the end of the school day. I included a required reading comprehension question to encourage students to do the reading before Friday's class. I figured if students read the article beforehand, they would have time to think about the points that interested them, why it was important, and what questions they could ask in the discussion.

The instructions I provided to the discussion leaders were simple. Start by giving a brief introduction to the article, including its main point and why you chose it, then begin the discussion by asking an open-ended question. Presenters were expected to prepare three to five questions, so when conversation after the first question quieted down, they could drive the discussion further with another question.

These discussions took place as a whole class. I know that it can be very intimidating for some students to speak in front of the class, but I hoped that by providing a variety of options for participation in addition to well-developed scaffolds (see Figure 2.3), even my shyest students would have the tools that they needed. The goal was to help even the most anxious performers reach a place of tentative comfort in speaking about science. For example, I made it clear that participation could be as simple as choosing one important sentence from the article and reading it out loud. I also provided students with sentence stems like "What interested me most was ___ because ..." and "This article relates to my own life because ..." so that they knew what kinds of comments contributed to a strong discussion. I even printed out "score cards" with point values assigned to each level of participation so that students could self-assess throughout the activity. These resources were available in English, Spanish, and Arabic, because those are the languages my students spoke. With the whole class engaged in discussion, there should have been plenty of room for conversation for 20 minutes, especially as students replied to each other and built on what their peers said.

Before I get into how this assignment actually went in class, I want to step back and tell you a little bit more about why I decided class conversations were so important. Honestly, part of me hated this assignment. When I was in high school, I dreaded having to speak aloud in front of the class. I was afraid my teachers would tell me I was wrong, I was afraid peers would make fun of me, and I figured "If I know it in my head, why do I have to say it out loud too?" But the more I learn about learning, the more I come to understand the importance of conversation when navigating through new, challenging content, and developing an academic identity.

In traditional education, a teacher presents information through a reading or lecture and expects students to internalize it. In this method, students are not provided opportunities to grapple with the content, to make connections, and to apply it to their own experiences, which diminishes their ability to actively learn (Carpenter et al., 2004; Johnson,

Name: ______________________ Period: _____ Date: ____________

Earth Science
Current Events Discussion

Complete the table below by recording your thoughts and observations during the class discussion.

Things I want to say:
Things I heard during the discussion:
Questions I have:

Discussion Participation SCORECARD!
Below is the SCORECARD that your teachers will use to evaluate your participation during the discussion.

	Teacher Notes
Listening actively while others are talking. (1 point)	
Taking notes on what other people say. (1 point)	
Summarizing information in the article. (2 points)	
Making a connection to class content or personal life. (3 points)	
Asking an open-ended question that drives the discussion. (3 points)	

Figure 2.3 The graphic organizer and scorecard I created to guide students during their weekly current events discussions. Images developed by the author.

2016). However, when students are centered in the learning activity, and have opportunities to construct their own knowledge, they are much more likely to get something from the content (Anderson, 2007; Johnson, 2016).

Student conversations can be an amazing way to make this learning happen. When you talk through something with your peers, you work through the new information you're trying to process, and develop linguistic skills that allow you to experience the content in different ways (Echevarría et al., 2012; Goldsmith, 2013). You figure out how to use your own vocabulary to explain things, then make connections to academic language (Johnson et al., 2013). My hypothesis was that if you did this enough times in class, you might lean into the idea that you have been acting and speaking like a scientist all along. Now that you have an understanding of what I was hoping for, I'll tell you what really happened. As you might imagine, I can't really claim that things went according to plan.

How It Actually Went Down

The first challenge was that few students came to class on Friday having read the article. While this was unfortunate for me as a teacher with high expectations, the presenter was most directly affected, as few students had the information necessary to participate in a discussion. When this happened, I gave the class ten minutes to read and annotate the article. On the board I posted annotation guidelines: Underline the main point of each paragraph, circle words you don't know and use a class dictionary to look up their definitions, and write comments and questions in the margins.

Some students took this time seriously, because they genuinely forgot to read the article before class. As I walked around the classroom, I observed them highlighting sentences and writing notes in the margins. For instance, in an article about the future of climate change in America (Fountain, 2019), one student wrote: "Will my building get flooded?" Other students used the time to look at their phones or rest their heads on their desks. I tried to engage them by asking them to find one sentence—even one word!—that stood out to them. I will never forget when one student shoved an article about the Moon getting farther away from Earth (Aderin-Pocock, 2011) across his desk, looked me in the eye, and said "Nothing about this interests me."

What I Learned

This was when I realized my big mistake: Student choice really has to mean *student* choice. Sure, I let students pick articles that interested them, but so few of them wanted to read articles in the first place! I understand in hindsight that this application of choice was superficial, and did not provide my students with true agency. In the future, I need to have a conversation with the class prior to developing the assignment, so that I can understand what they want to learn and how they want to learn it. Even though this might have been more engaging than some, it was still one more assignment students had to do. I realized that up until that moment, I hadn't been listening to my students as well as I thought, and needed to make a shift to truly know who I was teaching (Love, 2019).

In some of my classes, there was nothing I came to dread more than Fridays (and when a teacher is dreading a Friday, you know it's bad!). From the presenters coming to class and asking "What am I supposed to do?", to being able to hear the sound of students *blink* as the presenters and I asked probing questions to evoke participation … Sometimes it was brutal. I learned very quickly that if a group of students does not want to talk, there is almost nothing I can do to make them talk. I had to figure out how to do a better job of finding something they wanted to talk about.

The Silver Linings

The results of my experiment were not all bad. Given my goal of encouraging students to feel comfortable talking about science, this activity was most successful in my class composed of English language learners. It was a big class of about 30 tenth and eleventh graders from the Caribbean, Mexico, Central and South America. Some students had lived in New York City their entire lives, but some had immigrated to the USA that summer.

Because I am fluent in Spanish, I was able to teach this class bilingually. Students had the freedom to find articles in English or Spanish, and were encouraged to discuss in whichever language they felt more comfortable. One week, we read "Ask an Amazon Expert: Why we can't afford to lose the rainforest" from National Geographic (Schleeter, n.d.), and the next was "La NASA crea un mapa con todos los exoplanetas descubiertos hasta ahora y es asombroso" from Univision Noticias (Blancas, 2019). I provided translated copies of the articles so that all students could read them.

Though some conversation was lost in translation as not all students could understand both English and Spanish, students took initiative to provide articles and relevant videos to share with the class. These discussions often lasted the whole period, with students sometimes standing up to shout questions and answers at each other. A speculative article about what would happen if the Earth stopped spinning (Carpinetti, 2018) led to a full-blown shouting match rich with new ideas, questions, and opinions. This lesson format clearly supported these students' desires to learn through active dialogue. By shifting from a teacher-centered classroom to student-driven conversations, I was able to break away from the colonialist styles of traditional education. By engaging with content through collaborative discussions, students were given the opportunity to process information in a way that aligned to their own cultures (Adjapong & Emdin, 2015).

Though I am counting these experiences as a plus, I understand that fast-paced, spirited debates could be intimidating and off-putting to my shyer students. But still, I did manage to find success with those who preferred to not speak in front of the class. I was occasionally able to facilitate their participation by finding them outside of class and highlighting one of their skills and then encouraging them to bring that skill to the discussion. By saying things along the lines of "I noticed that you ask really good questions during labs … it would be awesome to hear one of your questions in the current events discussion this week," I was able to help some students find enough confidence to add their voice to the conversation. These one-on-one interactions showed my students that I truly did care about them, and helped them recognize their own strengths. In turn, this subtle new confidence was the boost that some students needed to feel comfortable being heard in class.

Patterns in Student Participation

Through my dogged attempts to chase students down prior to the Friday discussions, individual participation slowly creeped up in all my classes. I kept track of participation by tallying which students spoke during the discussions, and how many times each student spoke. Though my record of how many students participated each week looked more like a sine curve than a straight line, every week or two I could count on at least one new voice. It was nice to see that some of my students were moved by the strength of the relationship we had developed, and were ready to rise up to the high expectations I set for them.

I conducted this experiment in three classes. My Period 1 and 2 classes were integrated co-teaching (ICT), in which I co-taught with a certified special education teacher. Period 1 included students classified as English language learners (ELL) and students who qualified for special education services, mixed with general population students. My Period 2 class included students who qualified for special education services and general population students. My Period 8 class consisted only of general population students. Though the number of discrete students participating in each week's discussion varied, there was an overall increase over time across all three classes (see Figure 2.4).

Student Participation in Current Events Discussion

● Pd1 (ELL + SpEd) ● Pd 2 (SpEd) ● Pd8 (GenEd)

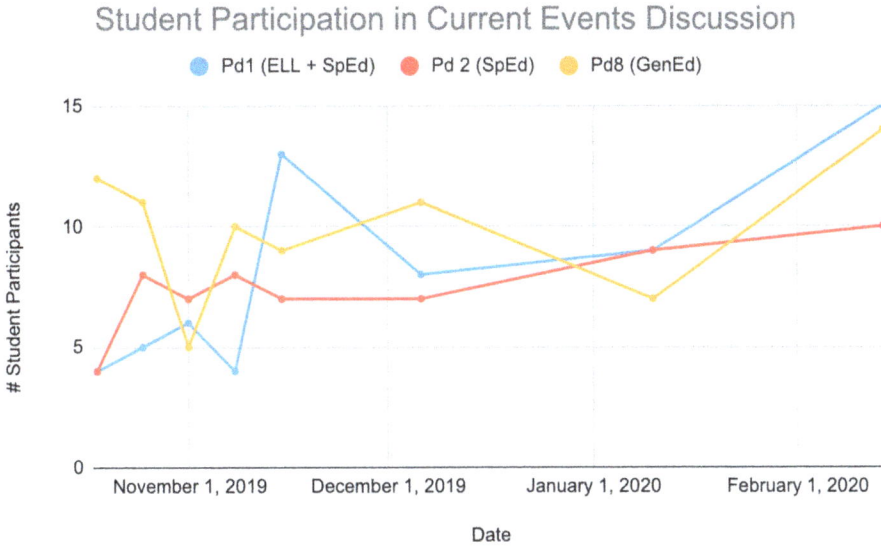

Figure 2.4 Graph showing the number of students who participated in each week's discussion in three classes and demonstrating an increase over time. Image created by the author.

Moving Forward: Teaching Science as Social Justice Inside and Outside the Classroom

As you can see, there were some positive outcomes of my experiment. Still, I'm not ready to call it an irrefutable success. I'm disappointed by how superficial the student choice aspect of this assignment turned out to be. I also realize that I fell short of my goal to offer a real scientific experience in my classroom where students were *doing* science—students were reading about science, but they weren't actively engaged in it. I thought that reading articles about important discoveries would help my students see how much of a role science can play in their own lives, but I think reading about it perpetuated ideas that "science is done by other people" and "science isn't for me." Why simply read about polluted groundwater when you could work in teams to design a solution grounded in science and engineering practices (NRC, 2012)? Why read about scientists searching for new exoplanets when you could design a plan to build a society on a new Earth? This is where I'm hoping to take my curriculum this year. Moving forward, I want my students to have the opportunity to grow comfortable talking about science through the process of *doing* science.

This is also an opportunity to involve my students in that critical stance of using science as a tool for social justice (Morales-Doyle, 2017).

When my students read an article about Greta Thunberg's role in the climate justice movement (Kwai, 2019), some were inspired to learn more about the ways climate change will affect their own community. I can take this a step farther by leveraging my curriculum to help students learn about the ways in which climate change exacerbates socioeconomic inequities, and challenge them to find solutions.

This Year Will Be Different

Dante, you'll see that this year we are doing things differently in our class. As you walked toward your seat on that first day of school, did you notice my new bulletin board? In letters as big as I could fit on a sheet of printer paper, it says "SCIENCE IS FOR EVERYONE!" and those words are surrounded by photographs of real scientists (see Figure 2.5). If you look closely, you'll see that scientists can fall on every part of the spectrum of race, gender, and ability. Give it a few weeks and you'll even see a picture of yourself and your classmates up there, because you too are scientists.

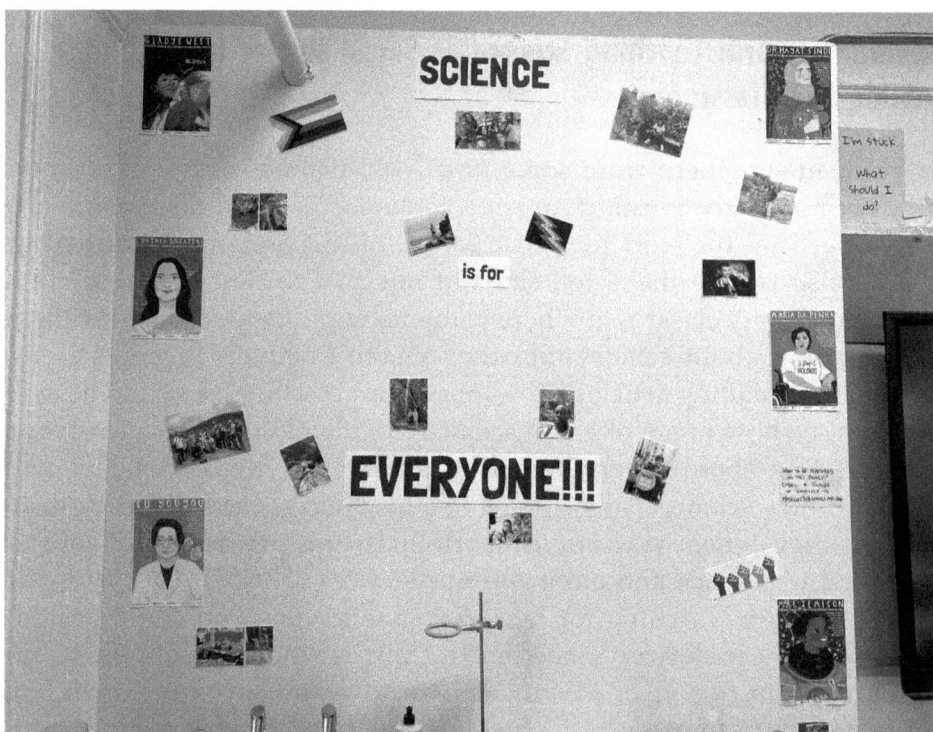

Figure 2.5 The "Science is for everyone" bulletin that I made at the start of the year. It features a diverse group of scientists, including photos of teachers at my school doing science. Photo and poster created by the author.

Over the course of this year, you will have many opportunities to do what scientists do. In class you will take measurements, compare your data with that of your classmates, create models that will help you make sense of the universe, and design solutions to pressing problems facing our society today (NRC, 2012). What I really hope you'll take away from this is that your role as a scientist does not end when you leave my classroom. Every time you try a new route to get home from school faster, every time you add slightly different amounts of seasoning to the meal you cook, every time you make a mistake and then try again: You Are Doing Science. And every time you do that, you are doing something to help make the world a little better, by adding your knowledge to the communal well-being.

Reflect on how you feel when you think about science as something that, in your words, "just isn't for you." Do you want your little brother to feel that way about science too? Do you want your children to feel alienated from science for their whole 12 years of school? If you answered no to that, think about what needs to happen so that they can rise above the biases of the system we are fighting against. Your teachers will help. We can start by ensuring that our classrooms are a warm and welcoming environment for you (NYSED, 2019b). Research demonstrates that students must first feel safe in a space before any profound academic learning can occur (Maslow, 1943). We need to be kind, respectful, and receptive to your needs. We must also make it clear that you belong in our classroom. We can start to do this by highlighting the work of scientists who look like you, and celebrating the historic and current achievements of diverse scientists, so you can see that "scientist" can be part of your identity too (Barakat, 2022).

We must provide you with as many opportunities as possible to participate in doing science. This involves introducing you to professional scientists, so that you can meet them and see for yourself that science isn't so "other" after all, while having organic conversations (Woods-Townsend et al., 2016). We can use class time to analyze the issues that are currently affecting our community, and apply what we know about science to seek solutions. We can support you in learning how to take a critical stance, using science as a tool to combat socioeconomic inequities (Morales-Doyle, 2017). I know this is a big responsibility. It's a lot to ask of you. But I also know that I will be by your side the entire time.

Most importantly, we must work together to demand that access to culturally responsive and sustaining science education continues to improve in our country. We must do everything in our power to make sure that we do not go back to the way things were (Love, 2020). It's time to retire the teacher-centered classroom, and use CRSE to include and empower all learners.

A Call to Action

Making a choice to say *"I am a scientist!"* might be scary. Inevitably you are going to make mistakes. You are going to fail. But that's what scientists do. Do you think that Mae Jemison was ready to operate a space ship the day she interviewed with NASA, or that Neil DeGrasse Tyson is able to film an episode of *Cosmos* in one take? Absolutely not. What has allowed them to be so successful is their ability to rebound from those pitfalls.

The next time you walk into our classroom, instead of saying "science isn't for me," try telling yourself "despite all my previous experiences, I am going to claim this space as mine." Think about how much power lies in that choice. Now follow that thought, extrapolate it, and take ownership in more and more places that you deserve.

Maybe right now you're a classroom scientist. But next year? In ten years? By the time you retire? Maybe you're the one who will use science to save the world. I know you will be.

> **Reflection and Discussion Questions**
> 1 Think about your students. Do any of them currently have a "scientist identity"? How might you find out? If they do have a scientist identity, what experiences (inside or outside your classroom) helped foster that identity? If not, why not?
> 2 What kinds of science experiences do you want your students to have in your classroom?
> 3 What can you do to learn more about any associations your students might have with science? What could you imagine doing to address any of the negative associations?
> 4 What structures, routines, and activities can you implement in the classroom to support all learners in feeling comfortable expressing themselves?

References

Aderin-Pocock, M. (2011, February 1). *Why the Moon is getting further away from Earth*. BBC News. https://www.bbc.com/news/science-environment-12311119

Adjapong, E. S., & Emdin, C. (2015). Implementing hip-hop pedagogy in the urban science classroom. *Journal of Urban Learning, Teaching, and Research, 11*, 66–77.

Almanac. (n.d.). *Full Moon for March 2022*. Retrieved January 3, 2022, from https://almanac.com/content/full-moon-march.

American Indian Alaska Native Tourism Association. (n.d.) *Native American Moon Names*. Retrieved from https://www.aianta.org/native-american-moon-names/.

Anderson, R. (2007). Being a mathematics learner: Four faces of identity. *The Mathematics Educator*, 17(1), 7–14.

Baines, J., Tisdale, C., & Long, S. (2018). *"We've been doing it your way long enough": Choosing the culturally responsive classroom*. Teachers College Press.

Barakat, R. (2022). Science and representation: examining the role of supplementary STEM education in elementary school student science identity. *SN Social Sciences*, 2, Article 25. 10.1007/s43545-022-00327-6

Baroutsis, A., McGregor, G., & Mills, M. (2016). Pedagogic voice: Student voice in teaching and engagement pedagogies. *Pedagogy, Culture, & Society*, 24(1), 123–140. 10.1080/1468136 6.2015.1087044

Blancas, E. (2019, November 15). *La NASA crea un mapa con todos los exoplanetas descubiertos hasta ahora y es asombroso*. Univision Noticias. https://www.univision.com/explora/la-nasa-crea-un-mapa-con-todos-los-exoplanetas-descubiertos-hasta-ahora-y-es-asombroso.

Brown, M. R. (2007). Educating all students: Creating culturally responsive teachers, classrooms, and schools. *Intervention in School and Clinic*, 43(1), 57–27. 10.1177/1 0534512070430010801

Campion, N. (2008). *A history of Western astrology. Vol. 1: The ancient world*. Continuum.

Carpenter, T. P., Dossey, J. A., & Koehler, J. L. (Eds.). (2004). *Classics in mathematics education research*. National Council of Teachers of Mathematics.

Carpineti, A. (2018, January 21). *What would happen if the Earth suddenly stopped spinning?* IFLScience. https://www.iflscience.com/what-would-happen-if-the-earth-suddenly-stopped-spinning-45340

Díaz, J. (2008). *The brief wondrous life of Oscar Wao (Pulitzer Prize Winner)*. Penguin.

Dutt, K. (2020). Race and racism in geosciences. *Nature Geoscience*, 13, 2–3. 10.1038/s41561-01 9-0519-z

Echevarría, J., Vogt, M., & Short, D. J. (2012). *Making content comprehensible for English language learners: The SIOP method* (4th ed.). Pearson Education, Inc.

Epstein, T., Mayorga, E., & Nelson, J. (2011). Teaching about race in an urban history class: The effects of culturally responsive teaching. Journal of Social Studies Research, 35(1), 2–21.

Falagas, M. E., Zarkadoulia, E. A. & Samonis, G. (2006), Arab science in the golden age (750–1258 C.E.) and today. *The FASEB Journal*, 20, 1581–1586. 10.1096/fj.06-0803ufm

Fields, C. D. (1998). Black scientists: A history of exclusion. *Black Issues in Higher Education*, 15(3), 12–16.

Finson, K. D. (2002). Drawing a scientist: What we do and do not know after fifty years of drawings. *School Science and Mathematics*, 102(7), 335–345. 10.1111/j.1949-8594.2002.tb1 8217.x

Fountain, H. (2019, December 4). Climate change is accelerating, bringing world 'dangerously close' to irreversible change. *The New York Times*. https://www.nytimes.com/2019/12/04/climate/climate-change-acceleration.html

Gay, G. (2018). *Culturally responsive teaching: Theory, research, and practice* (3rd ed.). Teachers College Press.

Goldberg, E. (2019, December 23). Earth science has a whiteness problem. *The New York Times.* https://www.nytimes.com/2019/12/23/science/earth-science-diversity-education.html

Goldsmith, W. (2013). Enhancing classroom conversation for all students. *Phi Delta Kappan, 94*(7), 48–52. 10.1177/003172171309400716

Hammond, Z. (2014). *Culturally responsive teaching and the brain: Promoting authentic engagement and rigor among culturally and linguistically diverse students.* Corwin Press.

Hannah-Jones, N. (2019). The 1619 Project. *The New York Times Magazine.* Retrieved from https://www.nytimes.com/interactive/2019/08/14/magazine/1619-america-slavery.html

Howes, E., & Wallace, J. (2022, August 9). Exploring culturally responsive science teaching through turbulence and challenge: Starting a multi-year research study during the pandemic. *AAAS Advancing Research & Innovation in the STEM Education of Preservice Teachers in High-Need School Districts (ARISE).* https://aaas-arise.org/2022/08/09/exploring-culturally-responsive-science-teaching-through-turbulence-and-challenge-starting-a-multi-year-research-study-during-the-pandemic/

Ilmi, A. A. (2011). The white gaze vs. the black soul. *Race, gender, & class, 18*(3/4), 217–229.

Irvine, J. J., & Armento, B. J. (2001). *Culturally responsive teaching: Lesson planning for elementary and middle grades.* McGraw-Hill.

Johnson, P. (2016 May). The role of conversation in the classroom—Promoting student voice through instructional dialogue. *IDRA Newsletter.* https://www.idra.org/resource-center/the-role-of-conversation-in-the-classroom/

Johnson, P., Betancourt, V., Villarreal, A., & Rodriguez, R. (2013). *Synthesis of effective teaching strategies and practices—A handbook for secondary mathematics and science teachers.* Intercultural Development Research Association.

Kwai, I. (2019, September 27). Are the kids alright in the era of climate change? *The New York Times.* https://www.nytimes.com/2019/09/27/world/australia/climate-change-youth.html

Ladson-Billings, G. (1995). But that's just good teaching! The case for culturally relevant pedagogy. *Theory Into Practice, 34*(3), 159–165. 10.1080/00405849509543675

Ladson-Billings, G. (2001). *Crossing over to Canaan: The journey of new teachers in diverse classrooms.* Jossey-Bass.

Ladson-Billings, G. (2014). Culturally relevant pedagogy 2.0: aka the remix. *Harvard Educational Review, 84*(1) 74–84. 10.17763/haer.84.1.p2rj131485484751

López, F. A. (2017). Altering the trajectory of the self-fulfilling prophecy: Asset-based pedagogy and classroom dynamics. *Journal of Teacher Education, 68*(2), 193–212. 10.1177/0022487116685751

Love, B. L. (2019). Dear White teachers: You can't love your Black students if you don't know them. *Education Week, 38*(26), 522–523. www.edweek.org/teaching-learning/opinion-dear-white-teachers-you-cant-love-your-black-students-if-you-dont-know-them/2019/03

Love, B. L. (2020). Teachers, we cannot go back to the way things were. *Education Week, 29*. www. edweek.org/leadership/opinion-teachers-we-cannot-go-back-to-the-way-things-were/2020/04

Maslow, A. H. (1943). A theory of human motivation. *Psychological Review, 50*(4), 370–396. 10.1 037/h0054346

Mead, M., & Métraux, R. (Eds.). (1953). *The study of culture at a distance*. University of Chicago Press.

Medin, D. L., & Lee, C. D. (2012). *Diversity makes better science*. Association for Psychological Science. https://www.psychologicalscience.org/observer/diversity-makes%20better-science

Morales-Doyle, D. (2017). Justice-centered science pedagogy: A catalyst for academic achievement and social transformation. *Science Education, 101*(6), 1034–1060. 10.1002/sce.21305

Morrison, K. A. (2008). Democratic classrooms: Promises and challenges of student voice and choice, part one. *Educational Horizons, 87*(1), 50–60.

National Academies of Sciences, Engineering, and Medicine (NASEM). (2018). *How people learn II: Learners, contexts, and cultures*. The National Academies Press. https://nap. nationalacademies.org/catalog/24783/how-people-learn-ii-learners-contexts-and-cultures

National Center for Science and Engineering Statistics (NCSES). (2021, April 29). *Women, minorities, and persons with disabilities in science and engineering*. National Science Foundation, Directorate for Social, Behavioral and Economic Sciences, National Center for Science and Engineering Statistics. https://ncses.nsf.gov/pubs/nsf21321/report

National Geographic Staff. (2019, April 9). Full moon names, explained. *National Geographic*, https://www.nationalgeographic.co.uk/2019/02/full-moon-names-explained

National Research Council (NRC). (2012). *A framework for K–12 science education: practices, crosscutting concepts, and core ideas*. The National Academies Press.

New York City Department of Education (NYC DOE). (n.d.). *School quality snapshot 2020–2021*. Retrieved February 12, 2023, from https://tools.nycenet.edu/snapshot/2021/

New York State Education Department (NYSED). (2019a). *Educator diversity report*. http:// www.nysed.gov/common/nysed/files/programs/educator-quality/educator-diversity-report-december-2019.pdf

New York State Education Department (NYSED). (2019b). *Culturally responsive-sustaining education framework*. http://www.nysed.gov/crs/framework

Pararas-Carayannis, G. (1988). Tsunami warning system in the Pacific: An example of international cooperation. In M. I. El-Sabh & T. S. Murty (Eds.), *Natural and man-made hazards* (pp. 773–780). Springer. 10.1007/978-94-009-1433-9_52

Paris, D. (2012). Culturally sustaining pedagogy: A needed change in stance, terminology, and practice. *Educational Researcher, 41*(3), 93–97. 10.3102/0013189X12441244

Paris, D., & Alim, H. S. (eds.). (2017). *Culturally sustaining pedagogies: Teaching and learning for justice in a changing world*. Teachers College Press.

Rathbun, J. A., Diniega, S., Quick, L. C., Grinspoon, D. H., Hörst, S. M., Lakdawalla, E. S., Mandt, K. E., Milazzo, M., Piatek, J., Prockter, L. M., Rivera-Valentin, E. G., Rivkin, A. S., Thomas, C., Tiscareno, M. S., Turtle, E. P., Vertsei, J. A., & Zellner, N. (2018, March 19–23). *The planetary science workforce: Who is missing?* 49th Lunar and Planetary Science Conference, The Woodlands, TX, USA. https://www.hou.usra.edu/meetings/lpsc2018/pdf/2668.pdf

Reynolds, J., & Kendi, I. X. (2020). *Stamped: Racism, antiracism, and you*. Little, Brown and Company.

Rumelhart, D. (1980). Schemata: The building blocks of cognition. In: R. J. Spiro, B. C. Bruce & W. F. Brewer (Eds.), *Theoretical issues in reading comprehension* (pp. 33–58). Lawrence Erlbaum Associates.

Schleeter, R. (n.d.). *Ask an Amazon expert: Why we can't afford to lose the rainforest*. National Geographic. Retrieved March 15, 2019, from https://education.nationalgeographic.org/resource/ask-amazon-expert-why-we-cant-afford-lose-rain-forest

United Nations. (n.d.). *The Paris Agreement*. Retrieved August 16, 2020, from https://unfccc.int/process-and-meetings/the-paris-agreement/the-paris-agreement.

Villegas, A. M., & Lucas, T. (2002). Preparing culturally responsive teachers. *Journal of Teacher Education, 53*(1), 20–32. 10.1177/0022487102053001003

Wallace, J., Howes, E. V., Funk, A., Krepski, S., Pincus, M., Sylvester, S., Tsoi, K., Tully, C., Sharif, R., & Swift, S. (2022). Stories that teachers tell: Exploring culturally responsive science teaching. *Education Sciences, 12*(6), 401. 10.3390/educsci12060401

Williams, A. D. (2020). *Genesis begins again*. Atheneum/Caitlyn Dlouhy Books.

Woods-Townsend, K., Christodoulou, A., Rietdijk, W., Byrne, J., Griffiths, J. B., & Grace, M. M. (2016). Meet the scientist: The value of short interactions between scientists and students. *International Journal of Science Education, Part B, 6*(1), 89–113. 10.1080/21548455.2015.1016134

Yancy, G. (2013 September 1). Walking while black in the 'white gaze'. *The New York Times*. https://opinionator.blogs.nytimes.com/2013/09/01/walking-while-black-in-the-white-gaze

3

Pop Culturally Responsive Education

Incorporating Students' Interests into a Scripted Curriculum

Samantha Swift

HIGHLIGHTS

- In this chapter, Sam focuses on the CRSE tenet of connecting to students' cultures, lived experiences, and interests to explore: *How can I use pop culture and student-centered practices to support science instruction while following a scripted curriculum?*
- Sam describes strategies and approaches that she used in her science classroom and what she learned through them, including a small group activity to elicit student-generated questions on weather and climate as a pre-assessment.
- While delving into her inquiry, Sam looks inward to consider her own cultural heritage and ponders how parts of her identity intersect with those of her students as she explores what CRSE means in her science classroom.
- We hope you will take away from this chapter ideas that you can use to learn more about the interests and experiences of the students in your classroom, and consider ways you might incorporate them into your instruction.

DOI: 10.4324/9781003397977-5

Introduction and Context

This chapter is about my journey discovering culturally responsive education (CRE), trying to implement it in my science classroom, and what I have
found to be successful. This journey is important to me, especially during
times when we are expected to follow a scripted curriculum, prepare
students for a standardized test at the end of the year, and hit all of the
standards in a certain amount of time that never seems to be enough. It
seemed that integrating culturally responsive elements into my science
instruction would be an impossible task, especially in a very diverse city
where you can have students with different backgrounds, cultures, and
experiences all in the same classroom. Diving deeper into elements of CRE
and the lives and interests of my students, it became clear to me that one
thing that many of my students have in common is the current popular
(pop) culture trends that captivate them outside of school. When I took a
deeper look, I discovered that pop culture trends are more than TV shows
they watch and video games they play in their free time—they are assets
that my students bring to the classroom and can be used in instruction. The
question I explore in this study is: *How can I use pop culture and student-
centered practices to support science instruction while following a scripted
curriculum?* I use multiple data sources to explore this question, including
instructional materials, student work, student attendance, surveys, and my
reflections as a teacher.

My Introduction to Culturally Responsive Education

I joined the Culturally Responsive and Sustaining Education Professional
Learning Group (CRSE PLG) in the unforgettable month of May 2020, which
marked the third month of pandemic lockdown, the death of George Floyd,
and massive protests across the country. At this time, I was living in Los
Angeles and teaching eighth-grade science remotely. More than ever, it felt
that I needed to focus less on academics, and more on my students'
experiences as they were navigating this new normal and watching protests
in their city. During this school year, this often looked like checking in with
students and their emotional needs at the time, or even talking about what
was going on in the world rather than front loading science content through
a slides presentation. After joining the CRSE PLG, I began to look through
the collection of literature that the group had assembled to find a definition
of CRE, and figure out how to apply it in my middle-school science
classroom. It was not an obvious connection for me—how to tie students'

cultural experiences into a science classroom—and possibly a bigger challenge to do it all virtually.

The following summer, I moved back to New York City (NYC) and dove into more literature about CRSE. I was searching for a solid definition that I could take with me from teaching in one diverse city, Los Angeles, to another on the opposite side of the country. While culturally responsive teaching (CRT) is defined in many different ways, one definition I read that really stuck with me was from Geneva Gay. Gay defined CRT as:

> Using the cultural characteristics, experiences, and perspectives of ethnically diverse students as conduits for teaching them more effectively. It is based on the assumption that when academic knowledge and skills are situated within the lived experiences and frames of reference of students, they are more personally meaningful, have a higher interest appeal, and are learned more easily and thoroughly. (2002, p. 106)

To me, this means making the content as relevant as possible to students' lives and experiences. This definition takes the remarkably wide variety of CRSE techniques and puts it into a simple sentence. Figuring out what CRSE looks like in practice, however, specifically in a science class, would be a huge challenge. In such a diverse city with different students in every classroom, I realized this must be a challenge that every NYC teacher faces. This seems especially pertinent when the P–12 population in NYC is nearly 56% students of Color, and the majority of teachers do not match that (NYSED, 2019a). This should be even more reason to be applying CRSE concepts.

In NYC, I took a teaching position in Manhattan at a middle school for grades 5–8. The middle school was a part of a small charter school network with one other high school in the same neighborhood. The neighborhood demographics were 72% Hispanic (New York City Department of City Planning, 2011). At my school, the population of students was 93% Hispanic, 87% of students came from households considered economically disadvantaged, and 20% were classified as English language learners, as disclosed on Public School Review (publicschoolreview.com). After working at the school, I learned that the majority of the Hispanic students were from the Dominican Republic. The demographics of this school were drastically different from the middle and high school I had attended at a large zoned public school on Long Island. The makeup of the school that I attended is currently 75% white, 20% economically disadvantaged, and 0% English language learners, according to Public School Review. Somehow it seemed even less diverse when I attended a couple of decades ago.

After my second year teaching in this Manhattan neighborhood, I met up with some of my aunts, who are in their 60s, with whom I hadn't spoken in a while. When they heard where I worked, they were filled with nostalgia and began to tell me that the school is in the same neighborhood where my great-grandparents lived, and where my grandparents grew up and met each other. I was shocked, as the neighborhood make up is so different from my ethnicity, which is Irish. It turns out that the neighborhood where I teach used to be majority Irish immigrants. My great-grandparents had emigrated from Ireland to NYC and lived in an apartment four blocks away from my school. Not only was it close to my job, but I passed the exact apartment where they lived every single day on my walk to work. The apartment was on the second floor above the luncheonette, as they described and I later confirmed using records on ancestry.com. I was so surprised to hear this and couldn't stop thinking about it for days. My grandparents had walked these same streets, in a neighborhood that had completely changed. I had unknowingly taken a job in an area to which I had a deep connection.

This made me think about how diverse NYC is and how, like many urban areas, it is constantly changing. Another layer of CRE is that it must always keep up with changing populations and our schools' communities. There could never be one curriculum that fits all. It also made me think about the future of my students and the power of education. Many of them are immigrants themselves, or children of immigrants, who came to the United States in search of more opportunities. Their opportunities start in the classroom. And I am fortunate to get to help them in this search. Being the first person in my family to graduate from college, I have seen how education can lead to more opportunities, and I find that would be similar for my students as well. In a way, it felt like I was meant to be at this school working with my students. It really inspired me to continue my CRE work.

CRSE during Remote Teaching

When I was about to leave California to move back to NYC, I found out that the upcoming school year (2020–2021) would begin virtually. It also presented another challenge—incorporating CRSE with a group of students I would be meeting for the first time through my computer. My initial idea for getting to know the students was to administer surveys as a mechanism for students to share their interests and hobbies. Being a science teacher, I am always looking for ways to collect data, and a Google Forms survey would provide fast results of every student's response. In Google Forms, students can answer multiple choice and short written questions in a survey format and send their responses to me. This helped me to find out about my students' favorite sports, what they watched on TV, their favorite subjects, and what they do in their spare time. These data were helpful at times, but I did not use them much because they felt disconnected from the students without my knowing them. Having never met my students in person, and not seeing some of their faces if they kept their cameras off, it wasn't easy for me to connect, even with these survey responses.

Around this same time, I was taking an early adolescent developmental course online through Stony Brook University. I was really inspired by a chapter in *What Every Middle School Teacher Should Know* (Brown & Knowles, 2014) about using student-created questions to guide learning. In that chapter, the authors called for an end to mandated curricula in schools, and highlighted a few schools that used a student-generated curriculum. I began to think more about this and its connection to Gay's definition of CRT that has stuck with me about centering learning in students' experiences. What could be more centered around students' experiences than exploring the questions they have asked? Going into the school year knowing very little about my students, and never having met them in person, it seemed like the perfect place to start planning. I wanted to let student-generated questions guide what we would learn in class.

After a few months of figuring out the basics of remote teaching—making sure that my students were showing up to class, getting the support they and their families needed during the pandemic, and creating structures for our online classroom—it was time to implement student-generated questions. I was hoping that this approach would reveal more about my students and their experiences, and that it would be a fun way to begin a new unit. In NYC public middle schools, different science subjects, including physics, Earth science, chemistry, and biology, are covered each year. Schools are expected to cover a certain set of standards across these subjects from sixth to eighth grades, in order to prepare students for a state science exam at the end of eighth grade. So the first challenge I faced was that I had to guide the students' questions towards our next topic: weather and climate.

Student-Generated Questions: Weather and Climate

During the school year, I followed the Amplify Science curriculum,[1] which uses a research-based approach to engage students in learning about scientific phenomena and is aligned with the Next Generation Science Standards (NGSS) (NGSS Lead States, 2013). Following Amplify Science was an expectation of every teacher in my school, and we were supposed to teach students a specific set of standards so that they enter the next grade level with the appropriate information. In middle-school science, the sixth-, seventh-, and eighth-grade units are often interchangeable and can be taught in any of these grades. This actually would make it perfect to try and have student-generated questions each year on different topics.

On the first day of our weather and climate unit, I modified a protocol that I learned about during an Urban Advantage[2] teacher professional development session to construct an activity with student-generated questions. I had students break into small groups of three and create questions on Google Jamboard, an online platform where students can work together in real time to edit a virtual whiteboard using digital Sticky notes. I chose six photos about weather and climate, placed them on the Jamboard, and had students come up with questions or noticings about each image (see a sample of students' questions in Figure 3.1). The only prompt that I gave my students was to create questions or noticings on a Sticky notes based on a particular photo, and to attach the Sticky notes note to the photo they selected. Each student chose a color for their Sticky notes note and developed questions based on the image. The groups got to see all six images together and create their own questions at the same time, and each group had their own Jamboard page.

Student participation was relatively high for a breakout room activity online. In my experience of remote teaching, it was common for students to go into a breakout room, turn off their cameras, and let one or two kids do all the work. Without teacher supervision in each breakout room, it was also common for students to be off task. Seeing the majority of students participating in the

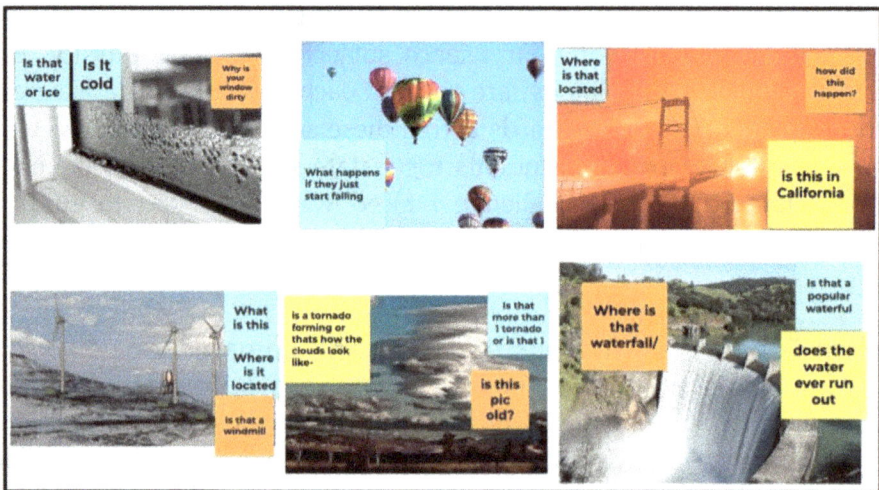

Figure 3.1 A screenshot of a Google Jamboard, a platform for students to work collaboratively. Screenshot by the author.

Jamboard meant engagement was high. I received great data from my students and got to work that night going through all of it. With 133 students on my roster and 90% attendance that day, I had a lot of Jamboards to sort through! I decided to download them as PDFs and organize them on my desktop.

Before developing a plan of how to organize the Jamboards and questions, I read through what students wrote. I found that this was a great way to access students' prior knowledge without giving some sort of formal baseline test. Three students worked together to create the Jamboard in Figure 3.1, where each Sticky notes color represents a different student.

In the first image with condensation on the window, I can see from the question ("Is that water or ice?") that the student needs to review the water cycle before this unit. This is a standard covered in fifth grade, where students should be able to recognize condensation, and know that it is liquid, not ice. I found this was one of the most common questions asked, indicating that students could not recognize that picture as liquid water on the window. Another question, "Is this in California?" reveals to me that this student has some sort of prior knowledge that California is often associated with wildfires. As this photo was taken the previous year, students may have seen it already on the news and associated the image with a place that was talked about often because of the forest fires. Finally, I can see that students in the bottom-right image recognize a waterfall, but there is no mention of a dam (this is a critical part of our next unit on weather patterns). The question "Does the water ever run out?" makes me imagine a whole unit based off of the question, "Will the water ever run out?" The unit could go in several directions—water conservation, the percentage of water available for human use versus in the ocean or frozen, or even a journey into the water cycle and how we have the same water here on Earth that was around when dinosaurs existed. Unfortunately, I knew exactly where this unit was intended to take us—explaining weather patterns, as it was part of the scripted curriculum I was expected to teach.

My first major take away was that this was a great pre-assessment activity. It revealed right away that we needed to review the water cycle before beginning the weather and climate unit. Students were looking at condensation and calling it precipitation, evaporation, or sometimes just asking what it was. We returned to the photos about the wildfires and lenticular clouds throughout the unit. I continued to use this activity as a pre-assessment for the remaining two units of the year. I chose photos that fit with the topic of each unit, and then looked at the questions students created as a pre-assessment. I was also able to continue this activity in person the following year, having students create Sticky notes and place them on poster paper on the classroom walls, rather than on a virtual

Jamboard. I found this approach way more engaging for the students than a multiple-choice pre-assessment. I noticed that students looked forward to the routine of opening a new unit by looking at photos, and that they seemed to enjoy the chance to write their own questions and read what others wrote. When this was done in person, they loved the chance to get up and walk around the room to place their Sticky notes on poster paper, and then read other posters with their classmates' questions.

Back in the Building and Getting to Know My Students in Person

The following year, all classes were back in the building. Being with students again in a classroom changed everything for me when it came to CRE. I realized that I only started looking for ways to implement CRE while teaching virtually and my perspective changed when I got into the classroom again. During the first week of school, I learned so much more about my students (maybe even more than my closest students online) and formed new relationships with them. Students and teachers were so thankful to be in person. I think that appreciation made it very easy to connect with each other. With this new social environment, it soon became very clear what students were interested in, based on how their backpacks were decorated, what they doodled in their free time, toys they brought to school, and what they talked about with their friends.

I never got to see any of this online. I had tried to incorporate social-emotional learning (SEL) questions into my virtual lessons, but nothing to the degree of being with the students for five classes a week, in person. Some of my attempts at incorporating SEL questions included lesson openers, such as "Would you rather go to the park or go to the arcade?" To my amazement, in one class, 100% of students responded by going to the arcade. It made me realize how little I knew them. I remembered that they are all kids, and they have a huge interest in games, TV, and pop culture in general. I define pop culture as entertainment, video games, sports, social media, and other forms of media that are consumed by a large number of people and aimed towards youth. Especially since their lives changed immensely over the pandemic, students were super in touch with the latest TV shows, popular video games, and Marvel movies. Some examples from this year were the video games *Fortnite*, *Among Us*, and *Minecraft*. TV shows that were popular among my students included *Demon Slayer*, *Naruto*, *Dragon Ball Z*, and *My Hero Academia*.

Even had I not given a survey to better get to know my students, what my students were interested in was clear. In particular, it was evident that anime was a huge part of my students' lives. From the first day of school, students were drawing and doodling anime, talking about *Naruto* (a popular anime show) and adding "anime" to the hobby section on their "about me" survey. (The "about me" survey was a homework assignment I gave to learn more about my students, asking questions including "What is your favorite subject?," "What is your favorite hobby?," and "When is your birthday?") I noticed that when students connected to our Google Classroom page, the majority used anime characters as their icons. When I walked around and asked students about the anime figures they were drawing, their responses sparked conversations. For the first time, I learned some of the most popular anime TV shows and characters without even watching the shows, but simply from listening to my students speak about them.

In fact, anime was so popular that students were really interested in Japanese culture in general, and many responded during an icebreaker question of "What place do you want to travel to?" with Japan. This shocked me a little, since 90% of my student population was from or had parents from the Dominican Republic (DR) so I anticipated that might be a common response. But it was clear from responses to this icebreaker question that they mostly wanted to learn about another place. I thought back to my early ideas about CRE, and before I even learned that term. To me, relating to students' background information was finding out about where they were from and making connections to that. But while many of my students had a connection to the DR, not all of them had been there in person or even had family living there anymore. Some of my students had really positive feelings about it, but not all of them. For example, I had one student who didn't enjoy spending time with his grandparents and dreaded going to the DR because he had to leave his mom behind in New York when he visited during the summer. He talked about it as if he would rather stay in NYC. It made me think that I could somehow begin to use this information about connections to places in my lessons, and reminded me to try and center learning around my students' experiences.

Using *Among Us* to Explore Claims, Evidence, and Reasoning

A popular online game at the time was *Among Us*. Almost every single one of my students had played this game. It is free to play and available on multiple platforms, which makes it extremely accessible to kids. I know my students liked it this year because I decorated my door with it and I had so many students commenting on it, asking if we were going to play in class. The reason I had decorated my door like this was because I remembered from remote learning last year that students loved the game.

A few times, when a small group of students showed up to my science class, we played *Among Us* online together. During this game, students had to accuse someone of being an "imposter" and back it up with evidence and reasoning. For example, in the game the characters run around a spaceship. In the spaceship they are all completing jobs, while the imposter kills off teammates secretly. After someone is killed, then the entire crew on the spaceship comes together to discuss who they think is the killer. In the game, one special characteristic of the imposter is that they can "vent" or travel from one side of the spaceship to another through the vent. Basically, if you catch someone venting, you know they are the imposter because they are the only player able to perform that function. This was thoroughly explained to me by my students using a claim, evidence, reasoning format in a casual conversation.

My students' explanations identifying the imposter in *Among Us* using a C-E-R scaffold

Claim: Red player is the imposter.
Evidence: I saw the red player venting.
Reasoning: Only the imposter is able to vent. Therefore, if the red player vented, it means they must be the imposter.

The conversation, where they explained this to me, was a notable teaching moment. Not only was I able to learn strategy in the game, but I had an insight when I asked the students: "What does venting mean? Why does that prove they are the imposter?" To help me understand, they were required to provide the reasoning behind their claims. This is what we strive for in science class when we have students connect their evidence to their claim. This connection or reasoning, in my experience, has always been the most difficult part for students in their writing of scientific explanations.

In explaining the game to me, students were having amazing discussions from student to student, stating a claim, backing it up with evidence, followed by reasoning. They were having so much fun, and were so passionate, making scientific explanations on their own without giving it the definition of scientific explanations. Knowing that constructing explanations is a science practice in NGSS and the new NYS science learning standards, I wanted to use the scaffold of claim, evidence, reasoning (CER) (McNeill & Krajcik, 2011). I thought about how I could incorporate it into my class. The reasoning is most commonly messed up and hard for students to learn, and I thought that this example might help them to develop a frame of reference, especially seeing their genuine interest in my door decoration.

Earlier in the year, at the end of my first unit on the microbiome, students were asked to write a scientific explanation to answer the question,

"How can a fecal transplant help to cure patient 23's microbiome?" Anyone familiar with Amplify Science will understand how the units provide students with the evidence they need to explain scientific questions, and teachers are left to guide the students on how to complete a full scientific explanation using a CER structure. In the curriculum, graphic organizers are provided for this, and the practice is explained using an example with Coca-Cola, which many students find helpful. In my class this year, when I was introducing CER, I decided to use *Among Us* as one of the examples.

We watched a short clip of the video game that showed an imposter "venting." I had set this up and screen recorded it with friends, in a very rehearsed game of *Among Us* to show exactly what I wanted. I then asked my students to make a claim identifying the imposter. They all raised their hands, dying to say "red," indicating that the imposter was the character in red. When I asked them how they knew, just as anticipated, they replied, "because red vented." When I asked what venting meant, they were able to articulate that it implied going down into the vent. When I asked, "How does that make them an imposter?," they explained that only the imposter is able to vent. From there, we transferred this thinking into a CER structure. This became a poster that I displayed in my room for some time (see Figure 3.2). The information they had just told me about the video game clip fit easily into a CER organizer.

In the past, I had tried to teach students about CER by asking the question, "Why is school closed?" This worked with some students, but not everyone. The *Among Us* example excited students, and I had much better reactions and products. It was easier to work with an example of CER that students already

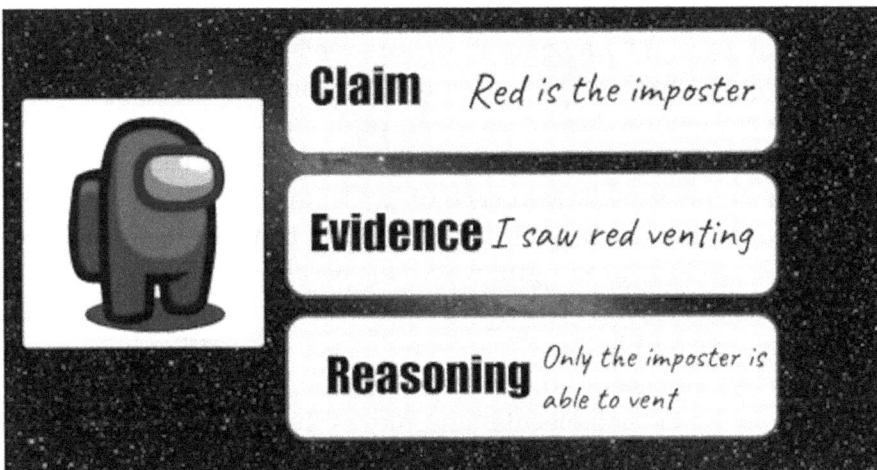

Figure 3.2 An artifact for learning in my classroom. This poster created on Google Slides acts as an anchor chart for when students are writing scientific explanations. It shows the character from *Among Us* on the left, and the example scenario we played out on the right in CER format. Artifact and screenshot by the author.

knew something about, using their experiences to guide their learning. This increased engagement and participation in this lesson. I could tell that, compared to my previous lessons, when students heard a reference to a video game, they paid attention. Because of the three-feet apart rule, which was a schoolwide COVID-19 safety measure at this point in the pandemic, students faced different directions in my classroom; when students were engaged, everyone turned toward the board, making their response and increased engagement physically visible. I am currently still evaluating CER in different forms, to see if my students are able to transfer using this structure to other situations in the science classroom, now that they had the *Among Us* frame of reference. For the rest of the year, we have investigations coming up where students will have to create a scientific explanation. I am anticipating great results!

Students had come into the classroom with assets and skills learned by playing this game. I believe that learning what students bring from their engagement in popular culture, and seeing these as assets, represents a CRSE approach (López, 2017; NYSED, 2019b). By uncovering these skills, I was able to use them to my advantage to help teach my students the science standards I was required to teach.

Incorporating Students' Common Interests to Teach about Cells

As I mentioned earlier, it was impossible to miss that anime was a common interest of the students in my middle-school classroom at that time. I often talked to other teachers in middle schools, and they shared similar stories and observations. Anime was not so popular when I was in middle school, and I did not grow up watching it. My students are always talking about anime, drawing pictures and dressing up for Halloween, to the point where I have watched shows based on their recommendations to try to stay relevant. With it clearly a huge part of my students' interests, I was always trying to use anime somehow in each of my lessons. Besides using pictures in slideshows or characters' names, I found that I couldn't really incorporate it into my lessons in a meaningful way. This became a struggle for me until one day a student asked if I had ever heard of the cell anime. As soon as she said it, I was amazed I had not discovered it yet. *Cells at Work!* is an anime series on Netflix dedicated to teaching about how cells in the body work, and how the body takes molecules and moves them around the circulatory system. With our cell unit about to end, I went home that night and tried to incorporate *Cells at Work!* into the remaining two weeks of my unit.

We started off by watching an episode from the series in class about the circulatory system. I then asked students to identify what in the TV show

Figure 3.3 A comic strip on cell movement created by a sixth-grade student. The caption states: "*Hello! Red blood cells are working hard, let's have a look. The red blood cells deliver oxygen to the cell.*" This is part of a page of her comic created for an assignment which asked students to write a story of how molecules are transported in the body to the places they need to go. In her assignment, she wrote a comic of how red blood cells deliver oxygen to all of the cells in the body. Photograph by author.

represented which part of the circulatory system. For example, the main character represents a red blood cell, the hallways represent blood vessels, etc. When I showed the episode in class, students were eager to get closer to the front of the room to see, and loved it. I then gave the students the option to write the next episode of the series to show cellular respiration (the topic we were reviewing at that time) or draw a comic (another thing my students love doing). For the assignment, I asked students to create a story of how different molecules get to the places they need to go inside the body. For example, one student created a comic that told a story of how red blood cells deliver oxygen from the lungs to all of the cells in the body (see Figure 3.3).

There was 100% engagement during this lesson. Students were asking for the name of the show, went home to watch it on their own, and told their friends about the science anime show. I had students from the previous year stop by my classroom to ask why they never got to watch it! The lesson was a win and was exactly what I was looking for. While I might not be able to incorporate a TV show into every lesson, now that they have seen these examples, we will continue to use them in class.

Reflection on Pop Culturally Responsive Teaching

Among Us and anime were foreign concepts to me before I started teaching sixth grade. I had to play games, reach out to friends, and sit with students to have them explain these things to me. At times I felt like they were teaching me—how to play *Among Us*, how to download it on my device, and start a game. Multiple students had a great time re-telling the plot of their favorite shows before I watched them for myself. But these are interests that my students care about, enjoy, and are relevant in their lives right now. These games and shows are already interesting to my students and they are a great jumping off point to teach the practices of science to my students. Figuring out what the newest show or game or even TikTok is requires building relationships with my students and finding out what interests them outside of school. The trends in pop culture are continually changing and require work to keep up with. Lessons that worked in 2021 might become irrelevant in the future. Students are entering the classroom with these assets and skills, and we need to do the work to uncover what they are.

What changed in my journey of incorporating CRSE elements into instruction was building relationships with the students and getting to know them better. Just being around students in person made all the difference. My first attempts at CRSE with my questioning strategies helped me learn more about their academic interests through a student-centered approach, but being in a room with students all day is really the way to get to know them. Talking with them, having them share stories or stop by during lunch, is how I found ways that engage them in science class by relating it to their experiences. This led me to discover a great way to introduce the science and engineering practice of constructing explanations. In the future, I would like to research ways to incorporate the other NGSS science and engineering practices, such as analyzing and interpreting data or engaging in argument from evidence, by using pop culture assets.

Considering my potential next steps for using pop culture assets: *Minecraft* in my Earth science classroom?

I have recently discovered the world of *Minecraft*, a video game that my students enjoy that contains a lot of references to Earth science such as certain rocks and minerals, and its ability to have students create worlds that can be used for assessments. There is a version that exists called *Minecraft Education*, seeing as so many educators have already picked up on its potential for the classroom.

Examples I've seen of using *Minecraft Education* in the classroom include:

- Creating different types of biomes
- Showing where water exists on Earth
- Creating a periodic table in *Minecraft*

Before the pandemic, trying to incorporate CRSE into my science lessons was difficult. I was in the mindset that scaffolding lessons to include CRSE elements into science instruction for each and every student with a different experience would be impossible to add to an already overwhelming workload. During the pandemic, I also faced the challenge of not being in the classroom with my students, and never getting to meet them in person. I was struggling to find out what the students wanted to know about, what they were interested in, mainly because building relationships virtually was tough. It was clear, coming back into the classroom, what my students were passionate about and what connected them to each other. Since I have started to incorporate pop culture into my class, I have seen engagement increase. Moving forward, I will continue to incorporate aspects of pop culture into my class whenever I can.

Because pop culture is always changing, this will require keeping up with my students in every class, making sure I know them, and modifying the way I teach. It would be much easier to use the same lessons and materials every year, and it would take less time to create them. But I think with CRSE, when teaching is centered around students' experiences, we need to recognize that those experiences are always changing, different for every student, and different in every class. However, many students do have pop culture in common, and putting the time in outside of school to keep up with the new trends is worth the engagement in the classroom.

Reflection and Discussion Questions

1   What assets or skills do your students bring to your classes?
2   What are students doing, watching, and playing outside of class that you can bring into the lessons?
3   How can you use these skills and interests to support your science instruction?

Prompt for journaling
Consider your own heritage. In what ways does your family and/or cultural background intersect with those of the students in your classroom?

Notes

1 From the Amplify Science website (https://amplify.com/programs/amplify-science): "Amplify Science is a K–8 science curriculum that blends hands-on investigations, literacy-rich activities, and interactive digital tools to empower students to think, read, write, and argue like real scientists and engineers" (Amplify Education Inc., 2023).

2 From https://bronxzoo.com/learn/professional-development/urban-advantage: "Urban Advantage (UA) is a standards-based partnership program designed to improve students' understanding of scientific inquiry through collaborations between urban public school systems and science cultural institutions such as zoos, botanical gardens, museums, and science centers."

References

Brown, D. F., & Knowles, T. (2014). *What every middle school teacher should know* (3rd ed.). Heinemann.

Gay, G. (2002). Preparing for culturally responsive teaching. *Journal of Teacher Education, 53*(2), 106–116. 10.1177/0022487102053002003

López, F. A. (2017). Altering the trajectory of the self-fulfilling prophecy: Asset-based pedagogy and classroom dynamics. *Journal of Teacher Education, 68*(2), 193–212. 10.1177/002248711 6685751

McNeill, K. L., & Krajcik, J. (2011). *Supporting grade 5–8 students in constructing explanations in science: The claim, evidence and reasoning framework for talk and writing.* Pearson.

NGSS Lead States. (2013). *Next generation science standards: For states, by states.* The National Academies Press. 10.17226/18290

New York City Department of City Planning. (2011). *Table PL-P3A NTA: Total population by mutually exclusive race and Hispanic origin, New York City tabulation areas, 2010.* https://www.nyc.gov/assets/planning/download/pdf/data-maps/nyc-population/census2010/t_pl_p3a_nta.pdf

New York State Education Department (NYSED). (2019a). *Educator diversity report.* http://www.nysed.gov/common/nysed/files/programs/educator-quality/educator-diversity-report-december-2019.pdf

New York State Education Department (NYSED). (2019b). *Culturally responsive-sustaining education framework.* http://www.nysed.gov/crs/framework

4

Helping Students Foster Emotional Connections

Connecting Students' Lives and Communities with the Natural World

Sean Krepski

HIGHLIGHTS

- In this chapter, Sean focuses on the CRSE tenets of connecting to students' cultures, lives, and communities and using students' assets as resources to explore the question: *How can I help foster students' emotional connections with the natural world?*
- Sean draws on object-based teaching and a passion for nature photography to develop learning activities and resources for his students, including a science autobiography based on experiences of being in nature and an ever-growing rock collection.
- Sean highlights how he was compelled to reassess his priorities as a teacher after the pandemic started, which spurred him to revise his approach and use the science content to make more connections with his students.
- We hope you will take away from this chapter ideas for visual and tactile approaches and object-based learning you might use to build connections between your students' lived experiences and the content you are teaching.

DOI: 10.4324/9781003397977-6

Introduction and Context

In the upheaval of teaching during the COVID-19 pandemic, I reevaluated my priorities as an educator. I teach tenth-grade Earth science at a public high school in Brooklyn, serving a predominantly Latinx population (71% of the students are Hispanic), many of whom are children of immigrants (NYC DOE, 2022). My school was created 15 years ago, after social workers fought to establish a high school in a neighborhood that had not had one for many years. The school is in close proximity to a beautiful historic cemetery with many examples of geologic features that I highlight in my class.

Prior to the pandemic, I focused heavily on teaching to the standardized test, which was emphasized in my school. I often lost sight of what I loved about science and nature, including things that can be entry points to get students interested. While teaching remotely during the 2020–2021 school year, I changed the way I teach my course to focus more on helping students foster an emotional connection with science and the natural world as a priority before focusing on content. The way I have redesigned my course centers on students' lived experiences and communities as entry points for teaching science. My work has been guided by the question: *How can I help foster students' emotional connections with the natural world?* In this chapter, I draw on teaching materials, student work, objects, images that I use in my classroom, and my own reflections.

Pandemic-Inspired Reconsideration of Teaching Priorities and Approaches

The pandemic and switch to remote teaching was, for me, a reset switch on my teaching. The whole situation was of course jarring and forced my colleagues and me to reassess our priorities and re-invent our science classes. This presented many challenges and difficulties, but there was a silver lining. I found in this experience an opportunity to make a better, fresher Earth science class. Prior to these changes, I was stuck in a rut with teaching Earth science solely to the standardized test, and I noticed that my students and I were often bored. Every year I would improve the class a little, but these were incremental changes.

With the disruption of the pandemic and online teaching, my colleagues and I had to rewrite the classes for a remote format. Working together, we were able to find approaches to teaching the material to better match our new priorities. After over a year of online teaching, we returned to the classroom, and I find that I am not interested in returning to the same old class. Instead, I am taking the organization and structure of our online class and using this to inform how I teach in person. My goal in this work is to help my students develop a sense of connection with nature and gain a deeper understanding of how their lives and communities are a part of it. For me, this connects with the culturally responsive and sustaining education (CRSE) tenets of making cultural connections by drawing on students' cultures and communities and using students' assets in instruction (Ladson-Billings, 1995; López, 2017).

Learning from Remote Teaching

I spent much of the 2020–2021 school year, when we were teaching remotely, reflecting on what I want my students to take away from my Earth science classes. At this point, my number one priority is to help my students cultivate an emotional connection with the natural world. I want to help spark their interest in science and nature by tapping into their lived experiences and connecting to their own communities. If students are to develop an intrinsic motivation to learn science (NRC, 2009; Teppo et al., 2021; Vansteenkiste et al., 2006), it helps to form that emotional connection with nature. This is important to me because I want my students to become lifelong learners and responsible global citizens with a greater appreciation and basic understanding of nature and their place in it. Years after taking my class, I know my students will not remember the details and minutia of everything they learned, or facts they memorized for a standardized test. My hope is that a basic appreciation for the world around them will be long lasting.

The change in my approach to teaching science remotely helped me to recenter my class around these goals. When we were back in the classroom, I took the lessons I learned during remote teaching, and the organization and style of the class, and adapted this approach to in-person teaching. As we once again must prepare students for standardized tests—something that was paused during the first years of the pandemic—I have changed my teaching to first access students' prior experiences with the natural world and then to build on them. This is important because who develops an interest in science or nature because they were taught facts and figures in a classroom?

Considering My Background

My new approach borrows from my experiences with informal learning environments. The groundwork for my own love of science was laid at a young age, not in the classroom but in museums and on family road trips, camping and hiking in national parks and summers spent out in nature. My own background is very different from that of my students, as I had a white suburban middle-class upbringing, and I teach a majority Latinx population and many of my students are children of immigrants. My students' lives and experiences are different from my own, and they come with their own associations and connections to nature. My job is to access these connections with different entry points and build on them. This is an important first step towards teaching science content: to anchor the content in students' lives and start establishing an emotional connection. This also aligns with the CRSE tenet of drawing on students' cultures, backgrounds, and strengths (Ladson-Billings, 1995, 2014; Paris & Alim, 2017). Throughout this chapter, I explore what I have been doing in my class to help promote these connections.

Science Autobiography Assignment

At the very beginning of the school year, before delving into content, the first assignment I give to my students is to write about their experiences with nature. My coworkers and I started doing this as an online assignment during remote teaching and I continue to use it. Students make a Google Slides presentation, called a Science Autobiography, in which they write about their favorite memory of science or being in nature. Occasionally a student will write about a previous science class, but much more often students write about experiences in nature. This assignment helps students reflect on their own experiences and becomes the major theme for the rest of the course, which centers on Earth science and how the students' lives and communities fit into it. The Science Autobiography also helps me understand where my students are coming from, in terms of their relationships with the natural world.

For this assignment, students have written about watching sunsets, leaving the city and seeing stars, visiting city parks, going to the ocean, hiking near the city, seeing wildlife, fall foliage, and the silence and peace of being out in nature. These are all entry points for teaching Earth science, helping students see its relevance, and begin to understand the larger world as the context in which their lives and communities exist. This is another place where my teaching practices connect to the CRSE tenets by valuing what students bring to the classroom as assets and using these assets for teaching (Ladson-Billings, 1995; López, 2017). After students share their lived experiences in nature, I can connect these to familiar local examples of natural phenomena. For instance, I can make comparisons to landscape features around the city, and incorporate local examples of weathering and erosion. I can then expand on these examples, generalizing these phenomena to the larger world. I can point out that the landscape features that students had experiences with were created by glaciers during the last ice age, and that weathering and erosion are processes that continue to shape the Earth's surface and our built environment.

A major theme that students bring up in these writing assignments is the silence and the serenity of being in nature. For example, one student wrote, "I walked through the forest and experienced the air and the quiet sounds." This is a quality I try to emulate in my classroom, making it a calming environment. I want students to feel comfortable and relaxed enough to focus (but not *too* relaxed, as there has to be a balance for students). I achieve this, in part, through soft lighting (an idea I got from the guidance counselors at my school), natural light as much as possible, and quiet, ambient music. One time my assistant principal came into my classroom while I was teaching and later told me it had reminded her of going to an aquarium—another type of informal learning environment—so I consider that a success. Often, students are confused at first about the lighting and music; but once they get used to it, many admit that they appreciate the change.

Once I have my serene classroom set up, have done my assessment to access students' lived experiences, and spent some time getting to know my students and building relationships, I use what I have learned to teach the Earth science content. For instance, when teaching about natural phenomena, I start with local examples from students' communities, based on what they write about at the beginning of the year. Teaching in a big city, building stones and tombstones in the nearby cemetery provide examples of rock types and weathering. Local hills, kettle ponds, and glacial erratics are excellent examples of a glacial landscape. I often share my own stories and photographs from exploring the city and encountering these features, and elicit stories from students in online discussion board assignments, prompting them to notice nature all around them. I also ask how weather variables (rain, humidity, air pressure, wind, etc.) affect their daily lives, and ask them to find rocks for our study of geology (see the next section on "Rocks from the Community"). After starting at the local level with things students have seen first-hand, I build on these examples to talk about how the same natural phenomena are at play in the wider world outside of the city. In this way, I want students to anchor their understanding of Earth science in something personal to them, see nature and Earth science in their own lives and communities, and think about how all our lives are part of a much bigger, interconnected system.

Rocks from the Community

Another new aspect of my class that my colleagues and I started during remote teaching is a Rocks from the Community assignment. For this assignment, students select a rock they like from their neighborhoods, take a picture of it, and develop a post for the online class discussion board. Students write a short description of the rock they selected and explain why they like it. In the past, my students have posted about pebbles and river stones that they found, amethyst crystals in their apartments, local building stones, and rocks from nearby parks. One of my students posted a picture of the first pebble they ever found on a beach when they were little and commented on the colors that they noticed on it. Another student shared a picture of a rock from Mexico, which was a gift from a friend who had taken a trip there (see Figure 4.1).

I have found this assignment to be a great jumping off point to talk about the rock cycle, and its ties to plate tectonics (which was already covered by this point in the course), because it is anchored in students' lives, memories, and experiences. In this way, I hope students will start to look closely at nature in their own communities and foster some personal connections to it. This approach borrows from object-based learning and informal science education (Kadar & Chatterjee, 2021; NRC, 2009), in that I use rocks, cobbles, and hand samples, including sandstone, in my class so that students can get hands-on experience with them. At times, I pair these rock samples with photos that portray the same rocks in their natural environments, so they become more than just chunks of rock in the classroom.

Additionally, it happens that students, completely unprompted, commonly bring small rocks to contribute to our classroom collection of rocks on display. These are examples that students can see up close and touch. I use these classroom samples in combination with photographs of what these rocks look like as part of a landscape to teach about geology. Among my teaching collection of rocks, I am now the proud owner of a small collection of pebbles and stones that students have brought me from their neighborhoods or from visits to their families' home countries, including Mexico, Ecuador, and the United States. (See Figures 4.2 and 4.3 for photos from the classroom rock collection.)

Sensory Activity to Build Connection to Natural Phenomena

Another strategy I use to build students' connections with natural phenomena is through sensory activities such as warm-ups at the start of the class period. For example, when I'm starting to teach about weather variables, I ask students to close their eyes and imagine a spring rainstorm. I ask them to pay attention

One of my interesting rock I have I one I found was at the beach. It's interesting because have some colors in it with scratchers. Also it was a first rock I had ever found when I was little on the beach.

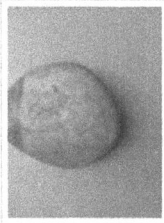

1- This rock came from Mexico lol 2- I thought this rock was pretty interesting because my friend brought it to me from her trip to mexico, plus its a tiny rock and thats how i like my rocks.

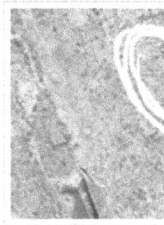

Figure 4.1 Two examples of responses to the Rocks from the Community assignment that students posted in the online discussion forum. Screenshots taken by the author.

Figure 4.2 Five pebbles that different students found and brought to me that have become part of my classroom rock collection. Included are pebbles from Mexico, Ecuador, and New Jersey. Quarter for scale. Photograph by the author.

to what it feels like, what it sounds like, what it smells like, and what it looks like. After two minutes, I ask students to share their answers to these questions. This activity can tap into students' memories, experiences, and imagination, and give them a sensory basis to start learning about weather phenomena. Students often share that they find the activity relaxing and that it is a good way to set a calm mood for the rest of class.

The first time I tried this warm-up was when I was teaching online early in the pandemic. Of course, in that setting, I could not actually see or hear the students, most of whom kept their cameras off. I set a timer for two minutes and afterwards a few of the students wrote their responses to the activity's prompts in the chat box. As with most of the online teaching experience, it was a little surreal and awkward.

Figure 4.3 An assortment of sedimentary rocks and fossils I have in my classroom. Photograph by the author.

The following school year, after returning to the classroom, I was nervous to try this activity in class and wasn't sure how well it would translate to the in-person setting. To my delight, students have enjoyed the activity and have given great responses. When called on, students most commonly have described the smell of the rain and the moist feel of the air. I use this as a basis to talk about weather variables like temperature, wind speed and direction, precipitation, humidity, and air pressure. The connections among temperature, humidity, air pressure, and rain is a great example. When the temperature is warmer, air pressure lowers so air rises, causing water vapor to condense, and rain to occur. These kinds of activities help students see the connection between their own lives, nature, and science. Perhaps most importantly, the two minutes of silence at the start of the class period sets a tranquil mood that then sets the tone for the whole lesson.

Photography in Earth Science Education

Visuals are such an important part of teaching about nature and science. Not only can images convey complex topics, but they can also help spark an emotional reaction, and provide context for the things I teach about in class. My approach to teaching leans heavily on photographs to root the content of the lesson in something students can see and imagine. In addition to being a teacher, I am also an avid nature photographer and I use my own photos as much as possible in my lessons. A photograph makes an object or phenomenon more real for the students than just a description or a diagram alone. Furthermore, it can be a powerful tool for visual storytelling and to convey ideas, access memories, and to establish an emotional connection.

In the first introductory lesson of the school year, I show several of my own photos, depicting rock, water, and trees. I use these photos to help elicit ideas from students about nature and what they're going to learn about in this class. I continue to use these and other photos throughout the semester, asking students to make observations and inferences about the ways in which rock, water, and plants may be interacting. Students become more engaged and ask questions when I tell them I took these pictures myself, and some students refer back to them later when working on their Science Autobiography assignment.

By using my own photography in my lessons, I make the examples more personal. This often leads to more in-depth conversations with students. They ask where I took that image, and how I took it. Often there's a personal story connected to the photo that I weave into the lesson. This frequently leads to deeper, more dynamic, and in-depth conversations with students, some of whom stay after class to continue talking. They want to know more about the places I visited and photographed, and to share their own experiences, which are often stories like those they wrote about in the Science Autobiography but are more informal and unprompted. This helps to make the object or phenomenon I'm teaching about connected to students' experiences and stories.

When I want to emphasize examples of Earth science in familiar places from the students' communities, I use photographs from New York City (NYC). For example, I use photos of the Hudson River when teaching about rivers (see Figure 4.4); outcrops of the Manhattan Schist (the bedrock in most of Manhattan) when teaching about rock types and NYC geology (see Figure 4.5); and glacially formed hills, erratic boulders, and kettle lakes in Brooklyn when teaching about glaciers (see Figure 4.6). This way, students can see real-life examples of Earth science in their own metaphorical, and sometimes literal, backyard! Other times, I use photos from my travels to zoom out and look at the wider world.

After I share my photos of nature, students are often excited to share their own photos, usually taken with their phones, of landscapes, sunsets, and travels to visit family. These informal interactions have inspired assignments. For example, during remote teaching, my fellow Earth science teachers and I made assignments for students to take and post photos of building stones, sunsets, and clouds. Beyond the assignments,

Figure 4.4 The Manhattan skyline, seen from Brooklyn. Here, the Hudson River and East River empty into Upper New York Bay. I use this photo of a local example when teaching about river systems. Photo by the author.

Figure 4.5 An outcrop of Manhattan Schist, the dominant bedrock of Manhattan, in Central Park, with skyscrapers in the background. Images like these help tell the story of New York's geology and geologic history. Photo by the author.

Figure 4.6 A glacially formed kettle pond, with glacially formed hills in the background, in a cemetery near the school. I show this photo in class and emphasize that this amazing example of glaciated landscape is a mere ten-minute walk from the school. Photograph by the author.

students are commonly eager and excited to show me their photos, unprompted, thus introducing me to their communities, friends, family visits, and sunsets.

Below are a few of my own photographs that I use in helping students see that Earth science is about the natural world and can help us understand why it looks the way it does (see Figures 4.7–4.12). Many of these are images from my travels, but I have included photos from the city where my students live to show how I connect their immediate environment to nature and to Earth science. Join me on a visual tour through my photo gallery for examples of photos I use to teach Earth science concepts, elicit students' ideas and experiences, and form connections with nature.

Photo Gallery

Figure 4.7 A pebble beach on the north shore of Lake Superior in Minnesota. This landscape was formed from igneous bedrock, solidified lava from a billion-year-old mid-continent rift. In the foreground is a metamorphic glacial erratic. This boulder is a different rock type than its surroundings, and it was carried here by a glacier during the last ice age, which ended approximately 10,000 years ago, the same age as the lake itself. I show this photo in class after students have learned about the rock types, weathering and erosion, and glacial features. I ask students to make observations and tell what they infer about the geologic history of this area. Photograph by the author.

Figure 4.8 Waterfalls spilling over a basalt landscape in northern Minnesota. I show photos like this to emphasize the interconnectedness of rock, water, and life, and how ancient volcanic eruptions made the foundation of modern ecosystems. Photograph by the author.

Figure 4.9 Sandstone cliffs in West Virginia's New River Gorge peak out of the trees with the bridge spanning across the river in the background. I use rocks, cobbles, and hand samples, including sandstone, in my class so that students can get hands-on experience with them, and pair these experiences with photos like this to show the same rock in the environment and the real world, not just chunks of rock in the classroom. Photograph by the author.

Figure 4.10 The Olympic Mountains of Washington State in Olympic National Park. When I teach about metamorphism in the rock cycle, I connect it to mountain-building and regional metamorphism. I use scenic photos like this one to inspire a sense of awe in nature, connected with the content students are learning in class. In addition to modern mountains, in class I teach about the Taconic Orogeny, a mountain-building event 440 million years ago that affected the East Coast. While those mountains have long since eroded away, photos of modern mountains can help students envision it. Photograph by the author.

Figure 4.11 A foggy day in a park in Brooklyn. I use photos like this to help students visualize weather phenomena, like fog, in their communities. This park is located near the school, and it is a location my students are familiar with, as many of them and their families spend time there. Photograph by the author.

Figure 4.12 A beach scene at Fort Tilden Beach in Queens. I use this photo as a visual when teaching about shoreline erosion and deposition, beaches, and sand dunes. Photograph by the author.

Conclusion

In this chapter, I explore multiple approaches that I took in my classroom to address this inquiry on *How can I help foster students' emotional connections with the natural world?* I began with setting the scene in my classroom. I use serene lighting and ambient music to do this. Next, I access students' prior experiences with nature using an autobiography assignment. Building on this, I also have students write about rocks from their community; find examples of Earth science all around them; and use local, familiar examples of Earth science phenomena that I can generalize to the larger world. I also use sensory activities to help students find a connection between themselves and nature. Finally, I use lots of visuals, including my own photography, to use natural settings and examples to establish personal connections, hopefully leading to deeper, more in-depth and meaningful interactions with students.

I believe that my current approach to teaching Earth science represents one way to implement culturally responsive and sustaining science teaching, as I use my students' experiences, ideas, and examples to ground my science teaching. In addition, I aim to use local examples to illustrate that even our urban environments show the results of forces and processes of the natural world. This approach has enriched my science teaching and I hope supports my students' developing knowledge about and connections to their local and global environments.

Reflection and Discussion Questions

1   In what ways do you bring connections to your students' neighborhoods and lived experiences into your teaching?
2   How might you consider incorporating local phenomena into your science teaching? What aspects of the natural world are near your school that you can consider bringing into your science teaching?
3   Are there ways in which you encourage and support your students to share about their own experiences with nature?
4   If you were to share photos or objects from your own lived experiences with your students, what would you share? What story would you use to accompany the photo or object? How could you connect this with your science instruction?

References

Kadar, T., & Chatterjee, H. (Eds.). (2021). *Object-based learning and well-being: Exploring material connections*. Routledge.

Ladson-Billings, G. (1995). Toward a theory of culturally relevant pedagogy. *American Educational Research Journal, 32*(3), 465–491. 10.3102/00028312032003465

Ladson-Billings, G. (2014). Culturally relevant pedagogy 2.0: a.k.a. the remix. Harvard Educational Review, *84*(1), 74–84.

López, F. A. (2017). Altering the trajectory of the self-fulfilling prophecy: Asset-based pedagogy and classroom dynamics. *Journal of Teacher Education, 68*(2), 193–212. 10.1177/002248711 6685751

National Research Council (NRC). (2009). *Learning science in informal environments: People, places, and pursuits*. National Academies Press.

New York City Department of Education (NYC DOE). (n.d.). *School quality snapshot 2021–2022*. Retrieved February 12, 2022, from https://tools.nycenet.edu/snapshot/2022

Paris, D., & Alim, H. S. (Eds.). (2017). *Culturally sustaining pedagogies: Teaching and learning for justice in a changing world*. Teachers College Press.

Teppo, M., Soobard, R., & Rannikmäe, M. (2021). A study comparing intrinsic motivation and opinions on learning science (Grades 6) and taking the international PISA test (Grade 9). *Education Sciences, 11*(1), 14. 10.3390/educsci11010014

Vansteenkiste, M., Lens, W., & Deci, E. L. (2006). Intrinsic versus extrinsic goal contents in self-determination theory: Another look at the quality of academic motivation. *Educational Psychologist, 41*(1), 19–31. 10.1207/s15326985ep4101_4

Part

2

CRSE and the Science Classroom

5

Behaviors without Behaviorism
Knowing the Students in Room 124

Raghida Nweiran

HIGHLIGHTS

- In this chapter, Raghida describes her classroom along with three archetypical students' classroom behaviors, wondering how to encourage those who need extra prompting and guidance from the teacher toward working independently.
- Through a student survey and reflections in her classroom, Raghida explores the question, *How can I learn more about the patterns of behavior that I notice in my classroom?* She draws on the CRSE tenets of valuing students' assets and holding her students to high expectations for their academic learning.
- Raghida reflects on what she has learned from her students in trying to understand these archetypes better, focusing on what they think teachers' and students' roles are in school.
- We hope you will take away from this chapter inspiration to think about your students' behaviors and what you might do to help them all succeed.

DOI: 10.4324/9781003397977-8

Introduction and Context

I'm a big rule follower. I love a detailed procedure, a good how-to diagram, maybe even an instructional video. During the summer when school is not in session, I brush up on techniques to use in the classroom to engage students, redirect their attention, and push their thinking. In certain circles, I'm the "planner" and in others I'm the "project manager," and these are not roles that I take lightly. I'd explain why, but this chapter isn't about me. It's about my classroom and my students.

I teach in a public charter school in a low-wealth and under-served setting in the Bronx. About 89% of my students are considered economically disadvantaged according to the state education department (NYSED, n.d.). The vast majority of my student population is Latinx and/or Black. Students flow through my charter network's elementary, middle, and high school, or they enter through a lottery when they are older. My school provides breakfast and lunch, a general counselor, a college and career counselor, Chromebooks and uniforms for all students, and other resources a financially stretched family might need.

In my seven years in the classroom, I've taught hundreds of students and interacted with many more. I have spent six of these years in the same school, where I have taught and continue to teach high school Earth science, physics, and computer science, from the introductory level to Regents (state exam preparation) to Advanced Placement (AP). My curriculum evolves annually in an attempt to keep up with the changing needs of my students. Each year is different, but the same, as new students remind me of old ones, colleagues leave and new ones come in, and school initiatives are recycled.

When principals, department heads, and donors come to observe my class, my students are on their best behavior. My students sit up straight. They understand that these important people dictate my day-to-day life and that, in turn, this will affect their experiences in my classroom. When administration is not around, my students are more human. Their behavior is not perfect and their curiosity shines through.

I have noticed that some of my students are just like me, but I have been puzzled by the others. I was curious about patterns that I noticed in some of my students' behaviors and wanted to learn more. Thus, I explore: *How can I learn more about the patterns of behavior that I notice in my classroom?* I engage in this inquiry through a student survey and my own reflections in my classroom.

Room 124

In my classroom, I strive for a warm and demanding environment. I'm friendly, but not a friend. I hold boundaries and give chances, and I show grace whenever possible. I expect efficiency and I know this influences my teaching, positively and negatively. I know how many questions my students should have answered in ten minutes, and I celebrate mastery. I do not recognize mediocrity and I expect my students to hold themselves to the same high standards that I know they can achieve. After all, holding students to high expectations is a tenet of culturally responsive and sustaining education (CRSE) (Ladson-Billings, 1995; Wallace et al., 2022).

I know my students can perform academically at high levels, as I have seen this in my science teaching. For example, in a laboratory class, my students tend to move a step forward toward that higher achievement working independently. I noticed that when they have lab materials to manipulate and a step-by-step procedure to follow, they were more likely to go ahead on their own than when the assignment was to read a passage or answer questions. Clearly, there is something more engaging for them in a laboratory-based lesson than in more traditional text-based classroom work. Seeing that they could be more engaged and succeed in laboratory activities pushed me to think more about what is going on in non-laboratory work.

In my teaching, I mess up just about every day, and through these mistakes and missteps I'm always looking to learn and to grow. I misspeak about content, and I have lost my patience with students before. I take ownership, apologize, and reset. From the literature, I know that students learn best when they are in a comfortable environment (Hammond, 2014; NYSED, 2019). I have learned from CRSE that if I can leverage the assets that students already bring with them to the classroom, then they can be intellectually challenged. The classroom learning environment has to include trust and the normalization of mistakes in order for students to feel capable of being vulnerable and taking intellectual chances. (See Chapter 7 on classroom environment.)

I share pieces of myself, like being white-presenting but traditionally Arab at home. I find out who has to translate for their parents, as I did as a child of immigrants. I never explicitly ask for details like this, but I share with my students what I feel is relevant at the moment. For instance, in the past I've commented at the start of class: "Oh, report card night? I used to be able to get away with paraphrasing, but my mom's English got better when I got older." Some students nodded knowingly, able to relate to my responsibilities of translating English for a family member. Their mothers also had figured out that "a little chatty, but gets her work done" actually means "talks all the time." I find that sharing details of myself casually throughout my lessons encourages my students to do the same.

Whenever I teach about volcanoes or igneous rocks in Earth science, I always think about my brother. When he was living in Hawaii, my family visited him and the volcano Mauna Loa. I show my students photographs of that volcano, as well as the rocks that we brought back from Hawaii. Students also bring in their experiences to share in class. When we talk about land and sea breezes, some of my students recalled and shared what the wind was like when they were on the beach in the Dominican Republic, for instance.

I spend the first few days at the start of school memorizing student names to show them that they are important to me. I know the names of my students' younger siblings, just like they know my little brother's name. I know which students walk home, and which students take the bus. Reciprocally, they know when I have to wait 14 minutes for the 2 train, and I know when the BX 19 bus is so full they have to wait for the next one. I know who has a job working after school and where. I am also aware of how that one student likes the red Doritos, but they pretend to like the purple bag because that's what everyone else likes. I know who struggles with reading and who stays for mathematics tutoring every week. I believe that my students know that I care about them. They know I'm silly, but serious about science, and that I will follow any tangent about outer space.

I conducted this inquiry in one class, in one discipline. Science is a historically white, male-dominated field. Students walk into my classroom with preconceptions about science as a discipline, or past experiences with science classes that can influence their performance (Gondwe & Longnecker, 2015; also see Chapter 2 in this book). I am upfront with my students about this issue. Teachers sometimes forefront images or stories of scientists of color to disrupt the white-male image that permeates the science field. I usually do otherwise, making it very personal and specific. I talk about and show my students where I went to university, in suburban Pennsylvania. My professors were predominantly white males; there were a handful of women, but no one darker than me. With my students, I scroll through the webpage of my university. Through this activity, centered on my own alma mater, I find that I can lean into this important discussion more, because I experienced it myself. When I ask my students for observations, they generally react with something like, "They're all white dudes." They let me know that this confirms what they already thought about scientists. I then challenge them with, "Okay, the way you change this is you become a professor. These people might not look like you, but that is all the more reason to get your voice heard in science."

The Students: Noticing Patterns and Archetypes

Each individual student brings their own personality, experiences, and insights to the classroom. However, I have noticed that across all of my students, there are a few common threads or archetypes that I see turning up regularly. I describe these here, with the understanding that no student fits any of these archetypes exactly. But these archetypes help me think about how I might improve my instruction for all of my students.

Archetypes of Student Behavior That I Explore in My Classroom

Student X: Student X is the student who has started classwork before I've finished going over instructions. Student X turns in completed work quickly, advocates for themselves, and asks questions only when necessary. I write report card comments such as "Student X is a model student." I bond easily with students who have the characteristics of Student X, as this is the type of student that I was.

Student Y: There is another student, Student Y, who begins, gets a little stuck, and then comes to me or a peer to get unstuck and progress. Some days Student Y is more like Student X, while other days Student Y takes a little break in between guided questions. Student Y generally turns in completed classwork and thoughtfully engages with the content.

Student Z: In all of my classrooms, I've also had another type of student, one who requires much more individualized attention. This student, Student Z, may not have a writing utensil, but will not ask for one. This student may sit quietly in front of their paper or laboratory materials, and, when prompted, will ask, "What are we doing?" This student will answer question one on the worksheet after it has been read aloud to them, but will not continue on to question two until they check to make sure question one was correct.

I've heard many terms to describe students who demonstrate behaviors like our Student Z archetype, such as "dependent learners" (Hammond, 2014) and "learned helplessness" (Maier & Seligman, 1976; van der Kolk, 2014). The behaviors I have observed are simple to describe and common to see. However, I have never understood why some students will sit and wait for a teacher to check in before trying something new, while others tackle the challenge without hesitation. All students are capable of learning and succeeding. Thus, I wonder, what motivates one group and discourages another?

I want to note that the behavior I'm describing is not specific to students with individualized education programs (IEPs). Specific accommodations, such as questions or directions read aloud, extended time to complete assessments, or the use of additional tools required for some students to learn, are not included in my observations of students such as Student Z. These accommodations are prescribed for some students and required for their learning. But I am not writing about students with IEPs here. I am thinking about those who wait for extra teacher input before beginning their class work, even if according to their academic record, this should not be necessary.

There are a couple of things that I have learned through conversation with others, and through my own experience, since I began this exploration. For instance, I know now that it is important to emphasize that a student resembling Student Z in my classroom may behave like a Student X in another class, right after mine, in the classroom next door (J. Lyiscott, personal communication on April 24, 2022). The archetype that a student may embody in my class or those associated behaviors do not necessarily indicate the same in another situation or context. A student who may need teacher feedback to proceed in my science classroom may be a self-starter in math, for instance. In my experience, I have seen a timid student with Student Y's characteristics in my classroom lead a school-wide town hall sponsored by the Muslim Students Association (MSA). As an advisor for the MSA, I saw this firsthand when I sat in the back of the classroom and one student resembling Student Y took charge and showed initiative. I admired their attention to detail, and their bravery when they spoke in front of nearly 100 people.

The rest of this chapter focuses on what I have learned from my students about their patterns of behavior and understandings of school, especially their role in the classroom. I implore you to not forget for a single second that I am speaking about multi-dimensional, complex human beings, and that each of their stories could never possibly be captured in a book, let alone a chapter.

What's Happening inside Their Minds?: Learning from a Student Survey

I personally relate to Student X's attitude toward school, so I set out to better understand Students Y and Z. I found particular interest in Student Z because my own experiences differed so much from theirs. My brothers were both like Student Y, which I found made it easier for me to relate.

To begin my inquiry, I hoped to learn what notions of school my students started with each day, so I could connect those assumptions with students' actual performance. My rationale going into this exploration was that students may have different perspectives of what school is or what it should be. I thought that students might express ideas about what it means to be a student, and what their role in a school is. I was hoping that this could lead me to better understand how dependent learner culture develops, and that this might potentially provide insights into how I could help them advance and gain confidence to become that self-starter. Listening to students and having them feel heard and valued is one aspect of culturally responsive and sustaining teaching that I wanted to convey through the survey I would administer.

I chose Google Forms to administer an anonymous survey, because I wanted my students to feel that they could express themselves openly, without fear of consequences. In total, I asked 103 students to participate, and received 72 responses. I intentionally asked open-ended questions on the survey, because I hoped students would be candid and explain their ideas. I asked questions such as, "What do you think you're supposed to do at school?" I didn't want a student to tell me what they thought I wanted to hear. I just wanted to see their thoughts.

The survey results showed me that my charter school students were very good at repeating one common idea. As I understand it, at some point in these students' academic careers, someone told them that students are expected to complete classwork, do homework, and study. Specifically, as the students wrote in the survey, students "should ask questions" and teachers "should give work and help students." Students thought that teachers are supposed to grade assignments and keep the classroom orderly. I received a clear message that teachers are supposed to teach, and students are supposed to learn. Students did not elaborate on how these things were supposed to happen.

I had the sense that students were telling me what they thought they were supposed to say, instead of sharing about their actual experiences. The survey responses reminded me of when a visitor would come to observe my classroom. My charter school students were used to donors touring their school. The minute an administrator or a stranger walked into the room, my students exhibited a well-practiced facade of trying to make themselves, me, and the school look good. They may know what they are expected to say or do as students, but how does this connect to what is driving the independent learners or what is stalling the dependent learners?

Reflecting and Regrouping: Back to the Drawing Board?

At my charter school, the majority of our students join us in elementary school and continue through our middle and high schools. I wonder how this trajectory affects my students, especially when compared to students in public school. I'd like to continue my inquiry into students' perceptions of the roles of students and teachers, even if it is with different students.

Surveying my students and writing about them was not an effort to place them in discrete boxes. The archetypes of Students X, Y, and Z capture a moment in a student's day. Students are amazingly multi-faceted, and those who do not excel in my classroom may be top performers in another. Even then, students who may not noticeably stand out, can shine and exhibit skills and talents in any number of ways. Taking this approach has allowed me to think about how I may draw on assets that my students exhibit outside of my classroom to engage them more fully in science learning.

I have learned about myself and my thinking about students through this study. But I want to learn more about students' perspectives about their learning. I wonder if I am asking the wrong questions. I don't want to lead the students to a particular answer, but I don't know how to prompt them to respond to what I'm looking for. In my survey, I asked "what" questions but I didn't ask "why." Should I rephrase and try again? Or am I missing something important? Students may share a common understanding of school and pull their motivation from other sources. I wonder what those sources could be.

What Happens Next?

A student brings everything to the classroom—their home life, history, personality, culture, and learned behaviors. In better understanding my students' thoughts and opinions about school, I can make their school experience more positive. School is more interesting when it is culturally responsive and sustaining. Natural curiosity and intrinsic motivation cannot be faked, but they can be encouraged. I feel that if I can better understand what is driving the independent learners or what is stalling the dependent learners, then I can modify my curriculum to better fit each student's needs. More specifically, I hope to continue conversations with students to fully honor their perspectives and cultures, and to encourage curiosity and other motivating factors in my science classroom. And continue to learn about my students as I teach.

In my classroom, I don't like connections between students and the content to be forced. I like these links to be based in my students' experiences. For example, when teaching about specific heat, I will say to students, "You only have ten minutes for lunch. Are you going to get a soup or sandwich?" Students will then share about their personal preferences for lunch. I may also say, "This is the food I like. What's the food you like?" I learn a lot about my students through these everyday connections to science. I still wonder what my students' personal backgrounds are like, including culture at home. I am curious about what their academic history includes, and if there have been any major shifts in their academic achievements or progress. These are questions that are more challenging to get at.

Student Z remains an enigma to me. I have developed classroom strategies to encourage Student Z to be more independent, like displaying a pen bin in the front of the room that they can access when needed. I also make sure to give directions both written and verbally, and leave them up on the whiteboard throughout class. I continue to wonder about Student Z and what they're bringing to the classroom. I will continue to provide all students with hands-on activities like the laboratory lesson described above, and strive to support all of my students including those with Student Z's characteristics in continuing to be successful in such settings. Culturally responsive and sustaining education rests on teachers' knowledge about their students. I know and continually learn a lot about my students. But I need to learn more about students' perspectives and motivations in my science classroom in order to be the best teacher I can for them.

Reflection and Discussion Questions

1   What behaviors have you noticed among your students that might fit into the archetypes named in this chapter? Are there different archetypes you notice in your own classroom?
2   How do the expectations and habits you have for yourself as a learner influence the expectations that you have for your students?

References

Gondwe, M., & Longnecker, N. (2015). Scientific and cultural knowledge in intercultural science education: Student perceptions of common ground. *Research in Science Education*, 45, 117–147. 10.1007/s11165-014-9416-z

Hammond, Z. (2014). *Culturally responsive teaching and the brain: Promoting authentic engagement and rigor among culturally and linguistically diverse students*. Corwin Press.

Ladson-Billings, G. (1995). But that's just good teaching! The case for culturally relevant pedagogy. *Theory Into Practice, 34*(3), 159–165. 10.1080/00405849509543675

Maier, S. F., & Seligman, M. E. (1976). Learned helplessness: Theory and evidence. *Journal of Experimental Psychology: General, 105*(1), 3–46. 10.1037/0096-3445.105.1.3

New York State Education Department (NYSED). (n.d.). *School report card.* Retrieved from data.nysed.gov

New York State Education Department (NYSED). (2019). *Culturally responsive-sustaining education framework.* http://www.nysed.gov/crs/framework

van der Kolk, B. (2014). *The body keeps the score: Brain, mind, and body in the healing of trauma.* Penguin Books.

Wallace, J., Howes, E. V., Funk, A., Krepski, S., Pincus, M., Sylvester, S., Tsoi, K., Tully, C., Sharif, R., & Swift, S. (2022). Stories that teachers tell: Exploring culturally responsive science teaching. *Education Sciences, 12*(6), 401. 10.3390/educsci12060401

6

Humanizing Science Teaching through Building Relationships

Kin Tsoi

HIGHLIGHTS

- In this chapter, Kin focuses on the CRSE tenets of drawing upon students' cultures, valuing what students bring to the classroom, and holding students to high expectations, to delve into the features of his classroom that seem to make it a welcoming environment. He asks the questions: *What are the features that make my classroom environment work? What are the "moves" that I make as a teacher that make the classroom more inviting?*
- Kin's in-depth reflection takes the reader on a journey through several different schools, analyzing what he has learned along the way about school cultures and developing a welcoming classroom environment that works for him and his students.
- Kin shares compelling stories about interactions and relationships with students, and includes results from surveys he used to learn about his students' perceptions of the science classroom.
- We hope you will take away from this chapter ideas for creating a learning environment grounded in conversation, honesty and trust building, and responsibility and opportunity; or possibilities for how you might explore features of your own classroom.

DOI: 10.4324/9781003397977-9

Dear Reader,

This chapter is written as a plea, not just to other teachers, but to myself: The classroom matters. In my experience, science and science class is often seen as cold, callous, focused on facts, accuracy, right and wrong, with test scores forefronted. Students come in afraid of being in science, and afraid of being wrong. But science is evolving and the context in which it is studied matters. We learn through mistakes and through experience. This chapter is a reminder that my science classroom should be a space to grow, learn, and exist; that nothing is so important in the curriculum that the students' learning and my learning about my students should not be taken into account.

Science is done by humans and far too often the human element of science is lacking in science teaching. I try to incorporate that element through banter and joviality and with the expectation that you will be wrong at some point and that's okay. For me, conversation is important, and learning the perspectives of the people in the room and trying to understand them is critical to both developing the curriculum and showing that you care.

Sincerely,
Kin

Context for Learning

I would like to start by saying how uncomfortable I am sharing things about myself. Creating a classroom where people feel valued is, for me, a LOT of work. To allow myself to be open has always made me feel vulnerable, which I do not enjoy. I am a teacher, male, Chinese, gamer, observer, roamer, foodie, and caffeine addict. I don't consider myself a researcher. I am primarily a teacher, working to make my classroom a welcoming and open space for students to learn and be themselves.

I have had the unfortunate experience of relocating to a different school three times in the course of my seven-year teaching career in New York City (NYC). As the least senior teacher, I was always the first one out. Through this experience, I learned a few things. The most important was to focus on the things that I can control, and for me that was my own classroom. In this chapter, I describe the lessons that I have taken from each of the school cultures I have experienced and how they have informed my teaching practice. I use the concepts of conversation, relationship-building, and classroom environment to illustrate what I have learned through my reflection about my varied teaching experiences.

My goal in this chapter is to share the story of my teaching journey, illustrated through journaling of anecdotal stories and conversations, reflections, and surveys. I've taught science in several schools and notice that something is working. This led me to consider what it is that is working. Hence, I explore the questions: *What are the features that make my classroom environment work? What are the "moves" that I make as a teacher that make the classroom more inviting?* I will show what I've intentionally sought to learn from my students, and incorporate their survey responses as data. And through all of this, I hope to share how I grew as a teacher and a person, and how my Earth science classroom became a space for people to succeed and what that means and looks like. It is important to note that the structure of my classroom works because of who I am and choose to be when I am there. It is specific to me; thus, no one should copy this format with the expectation that it will work. Kids, after all, can sense when something isn't true from a mile away.

I have come to understand that culturally responsive and sustaining education (CRSE) cannot be done via checklist. It is through engagement with the students and learning about them that I am able to have high expectations and build lessons that incorporate their experiences. In my teaching and inquiry, I explicitly address three CRSE tenets: drawing upon students' cultures (Ladson-Billings, 1995a, 2014; Paris & Alim, 2017), valuing what the students bring to the classroom (López, 2017), and holding them to high expectations (Ladson-Billings, 1995b) (see also Chapter 1). I want my students to feel that they are a part of the classroom community. This means that I must see each student as a person when designing my instruction to create multiple avenues for them to learn, and to help students believe that they will learn something and be able to prove it.

The Family Academy for Data Sciences:[1] A Lesson That School Should Be a Welcoming and Safe Environment for the Students

My teaching journey starts in a relatively small high school I call The Family Academy for Data Sciences. Most students at this school were testing below grade level. What the school did well was build itself up as a space for the students, leaning into the idea of school as a safe haven. It was a young school that had their first graduating class during the year that I started. By the admission of a senior staff member, the school gave too much leeway to the students who started in the school's first year. They were their babies. To some extent, it worked because they got students to buy into the school. But as the years went on, the structures that they created were not very effective. For example, "pride dollars," an incentive-based reward system, was effectively neglected in the two years that I worked there. From what I could tell, there were no tangible benefits to earning pride dollars except as a "good job" or pat-on-the back. It did not help that it was never clear what the value was for a pride dollar. By my second year in the school, pride dollars had been phased out.

What *did* work was the well-run advisory system, put in place at the inception of the school. Each teacher had a set of about ten students to supervise and monitor. Every day, students' advisory groups would meet and talk about the topic of the day or do an activity that the advisory committee created. Students were given an avenue to talk about their life experiences and had benchmarks to meet. Each advisory group had a name and a personality. It gave teachers a chance to get to know students and for them to know us, because we (the teachers) were expected to share about our lives as well. This experience informs my approach to building my classroom community.

The multiple school assemblies were another aspect of this school that supported the welcoming environment. These assemblies centered around the primarily Black student community. The school assemblies mirrored the local neighborhood community, which was very impressive, especially considering that CRSE was just starting to enter the NYC schooling culture. The guest speakers at the assemblies talked Black issues, they talked Black cultures, they shared Black stories. And in all of these, the students sat with their advisory groups and shared their own stories. Once the school invited a spoken-word poet who gave a clear message: *It doesn't matter what your grades are. It does not take away from your value as a person. But the system itself is not great and, in order for you to succeed, you have to work.* It was a humanizing experience, centering the students and their backgrounds, and applying it back into their current school environment. Having a successful Black poet show up talking about how they got to their position, citing the issues that they dealt with, validated students' concerns but also gave them hope and a path to follow.

In my opinion, the biggest failure in the school was the lunchroom. People did not want to be there. Instead, several students came to my room to sit, eat, and hang out with their friends and with me. I thought it was wonderful and I was all for it. For whatever reason (the administration decided that the teachers should not have a say in this), students were told they were no longer allowed to go to the classrooms during lunch. It was a shame because I had a small group of five boys who regularly stayed behind in my room, talking about video games and basketball. It was unfortunate that the administration put a stop to it. Nonetheless, the students loved the school, and just wanted to be there—even the students who struggled academically. They enjoyed after-school clubs, many of which focused on equity through a non-profit organization, and a handful of schoolwide events each year. From what I learned during my time there, many students felt that the school environment created a space of belonging and unity.

However, my first year of teaching Earth science at The Family Academy for Data Sciences was difficult. As a brand-new teacher, I had high ideals and I wanted to change the world. One of my goals was connecting everything that we learned in my science class to my students' lives, believing that was the best way for students to understand and appreciate the material. Earth science, by nature of being about the world they live in and exploring phenomena related to the students' lived experiences, provides a relatively easy route to mesh these together. Connecting back to the students and their world was all I wanted. We now call these cultural connections part of CRSE, with an understanding that culture is not restricted solely to where your family comes from (NYSED, 2019; Wallace et al., 2022).

Connor and the Rock: A Lesson in Creating a Safe Classroom Environment

One of the lessons I taught in my first year of teaching Earth science addressed the misconception that all rocks were created in situ and that they have never changed. One day, we were discussing how rocks break. I posed a question asking where the rock samples used in class came from. After a mixed set of responses from my students, as was expected, I took out a large sample of sandstone and asked them about the size of the sample relative to the usual rocks that we worked with in class. Then I took out a rock hammer. And in my infinite wisdom as a new teacher, I allowed a handful of students in each class to use it (safely) to break off a small sample.

It was around the middle of February that this lesson took place, so I had already spent half the school year with the students. I knew them reasonably well and trusted my understanding of what they could and could not do. I had brought in a rock hammer for them to:

1 See that smaller rock samples come from bigger rock samples,
2 See that a lot of the samples used in class were personally collected, and
3 Experience rock breaking.

The story I'm relating here took place in what was considered by other teachers in the school to be a challenging class academically and behaviorally. I believe that the issue was that nobody believed in this group of students. Ironically, that was my most well-behaved class that year. In the lesson, we had discussed various ways that rocks could be broken. I shared how I went with other teachers, as part of my teacher preparation program, to collect rock samples. Then, I offered some of my students a chance to help break a part of the bigger sandstone. I did not just want the best behaved or the highest achieving students to participate. I wanted a good mix because I felt that it was a powerful geology opportunity to offer them.

One of the students I chose, Connor, had a bit of a reputation. The year prior to my start in the school, he had allegedly threatened to stab his biology teacher. There were several other infractions and write-ups from other teachers. I truly think he just had a lot of energy. And I really do mean a LOT of energy. I remember him screeching at the top of his lungs, running around the two floors the school occupied and walking in and out of the classroom constantly. I like to think that in the end, I had a really good relationship with him and that it was mostly because of a mutual trust that we built in my time at this school. And I think that most of that came from the incident that I am about to describe.

As mentioned, I wanted a good mix of students to partake in breaking the rock, and in his class, Connor was the last student I chose. Some students were shocked at this decision. Connor was excited. He had a bunch of energy and finally had some way to get it out. So, I let him use the hammer. He did everything that I asked of him in order to make sure he was safe. Then he broke off a chunk of sandstone for me and that was it. Later in the class, Connor asked if he could hold onto the rock hammer for a bit at his desk. I was hesitant, but I also thought it was a good chance to build rapport with him so I let him. And Connor did great. He was more patient than usual; he got his work done; he sat down; he raised his hand; he asked questions. He was great.

I wish that were the end of that story. There were probably about ten minutes left in the period. One of the hallway staff, who ensured the flow of students between classrooms and was responsible for the bathroom keys, walked by the classroom and saw that Connor was holding the rock hammer. This particular staff member had heard the same stories about Connor that I had. When he saw that Connor was holding the rock hammer, he came into my classroom and raised his voice, trying to convey to me that Connor was dangerous.

"Connor cannot be allowed to hold the hammer. He is a danger to himself and everyone around him," he said. It's been five years, and I still remember those words clearly. More importantly, I remember Connor's response. I asked Connor for the hammer. He gave it to me. Connor was about to walk out of the room to go see his guidance counselor, someone whom he trusted and had worked with for over a year, to talk about what happened. At that point, the hallway staff decided to stop shouting at me and began shouting at Connor about his behavior. Connor was practicing restraint. I remember talking him down. I kept telling him, "Connor, you didn't do anything. You'll be fine. Right now, it's all on me. You're okay. Let me deal with it." I wanted to remind him that he was safe here, that he had done nothing wrong to that point, a position that I still stand by today. He did exactly as I asked.

But the hallway staff continued to shout at Connor. There was a point when Connor could no longer take it. He took the piece of sandstone that he had broken off and threw it at the hallway staff. It did not hit him, but at that point the situation had escalated. The other hallway staff were called, and the assistant principals became involved. The rock hammer was confiscated. Connor was suspended and visibly upset. He did not want to go. He hid in the closet and the guidance counselor had to coax him back out. I think he was disappointed in himself, and that he believed he could have done better.

This incident led to one of the most difficult phone calls that I have ever had to make. I could hear his father's exasperations and sighs. It was the sound of defeat, of "What did he do this time?" He was used to getting behavioral calls about Connor. I remember telling Connor's father that I was sorry to make the call and that Connor did really well until he threw the rock, and his father turned around a bit. I was not calling home because Connor had done something wrong or misbehaved—at least not in my view. He was provoked and he lashed out, like many of us would have done. How many people would stand there to be talked about like they were not even there, not relevant? How many people would stand there to be berated by someone for what probably felt like hours (but was in reality minutes)? Connor had understandably lost control of his emotions, despite trying so hard to maintain them.

So much about the situation could have been handled differently. I don't teach that lesson anymore. I don't bring a rock hammer to school or let students try to break my larger samples into smaller ones. They learn about rocks breaking through readings, videos, and less active means. It's a shame. I think it was a fun lesson and the kids enjoyed it. I think that it made the unit different. But, I can't take that risk anymore.

For me, this was a story of a failure. I failed because I couldn't stop Connor from throwing the rock. What could I have done? I think now that this might not have happened if my classroom had been a more safe space for my students. For me, that story ended when I spoke with the assistant principal about Connor's grade and asked, "How do you expect me to have high expectations and give opportunities when things like this happen?" and all I got was something akin to "I'm glad you asked that question."

The following year, I spent more time working on setting boundaries in my classroom. It was more rigid; it focused on developing specific skills in science, it made students work, and there were successes. But something felt like it was missing and I ended up changing schools.

The Melting Pot High School: A Lesson That Students' Assets Need to Be Valued

The Melting Pot High School is a medium-sized multicultural high school in New York City geared towards students who have arrived in the country within the past two years. It was very different from The Family Academy for Data Sciences. Surprisingly, in a school that was geared towards students from a wide range of countries, it tended to take a deficit perspective regarding the cultural assets of the students. My conversations with administrators and other teachers were either about what the students were "lacking," how the school was struggling to stay afloat due to challenges with enrollment, or students' "poor grades."

For instance, I recall a student who was switched to my class after the first week of school who was one of my best students that year. When she came into my class the first day and asked where she should sit, I told her wherever she felt comfortable and she seemed surprised and confused. She told me at the end of that day that the way that I teach and the way that I structure my class reminded her of a cousin—a note that I was, in turn, very confused about. She assured me that it was a good thing and that it was a nice change from a strict and rigid setup. Later in the year, she started having some life issues. I remember it being related to cancer although I don't recall if it was her or her mother.

I asked my AP for advice, with something along the lines of "Did you know about Izzy Rose?" because his response was "Yes, of course I know about Izzy Rose." And then he proceeded to rattle off some of her grades and mentioned that every year she starts off doing well and then her grades start slipping. At this point though, I already knew Izzy Rose to be a good student who was attentive and participated and tried. The Izzy Rose I knew stayed behind to work on assignments if she felt like she was slipping. The person my AP was describing was not the Izzy Rose I knew. The AP mentioned nothing about the actual life concerns that Izzy Rose was tackling. Before the school year ended, Izzy Rose dropped out of school. She stopped by on her last day and spent five minutes talking to me about what she wanted to do with the rest of her life.

As illustrated with Izzy Rose's story, at this school, everything was either about test scores, course grades, or student enrollment. In the beginning of the year, one of the staff meetings was about low test grades. Being new to the school, I went to talk to the other Earth science teacher. I remember the underlying message used in the conversation: "The kids are bad. They don't know anything." When I asked this teacher how they taught, the message was "I know how to teach, stop asking me so many questions," as if it were beneath her to help the students or to help me. In the time that I was there, there was little discussion about how we could integrate the cultural assets and the varied backgrounds and interests of the students, or even whether there *were* any student assets.

At Melting Pot, I felt that there was little effort to see the students as individual human beings, and there was a disregard for students' perspectives, what they knew, what they offered, who they were. In the entire school year, there were a total of two events that celebrated the diversity of the students. First, there was a multicultural food festival. Students from various nationalities shared foods from their cultures. I recall booths for El Salvador, Mexico, Colombia, Japan, Egypt, Korea, China, and Russia. This event was mentioned only once in a staff meeting, and I found out about it largely from students who were excited to share a part of themselves. I enjoyed my time at the festival with the students, trying the different cuisines. I remember students' looks of concern as I tried certain dishes as if waiting for my approval. I tried chicha, a fermented corn-based drink from some of my Colombian students; tacos dorados, a rolled taco, from the Mexican station; edamame and onigiri from my lone Japanese student; and grape leaves from my Egyptian students. I had fun learning about the students, where they came from, their cultures, and their habits.

But for the administration, it seemed like a box to be ticked, blank and emotionless, instead of a festive vibe and celebration of our students. Immediately after the food festival, we went back to the same tired conversation about "How can we improve test grades? How do we increase enrollment?" It was not until the Festival of Nations that the multilingual, multicultural aspects of the student body were forefronted again. However, as with the food festival, it was not well advertised. Izzy Rose, previously mentioned, was really excited to be in it. She shared with me that she was going to sing in the festival and wanted to make sure that I was going to be there. Obviously, I went. I remember dances and singing.

These types of activities—"foods and festivals"—have been described as superficial in terms of culture. But they were, for me, an opportunity to learn more about my students, interact with them outside of the classroom, and think about how I might connect my students' favorite foods and cultural traditions to my science teaching, making cultural connections in instruction.

Administering a Student Survey

Considering that I wanted to learn more about students' perspectives and who they were as people, I administered a survey to my three Earth science Regents classes.[2] The purpose of the survey was to see whether students thought they were represented and seen in this school and in the classroom. I received surveys from 52 out of approximately 80 students.

I provided four statements on the survey, and asked students to indicate their level of agreement with each using a five-point Likert scale ("1" indicated strongly disagree, "3" was neutral, and "5" strongly agree). The statements included:

"I feel like I can explain my group culture to someone."
"I feel like my group culture is represented in the school."
"I feel like my group culture is represented in this class."
"I feel like I am represented in this class."

In general, students' responses suggested that the school itself did not do a good job of incorporating all students. Nearly a quarter of my students (23%) indicated that they felt their group culture was underrepresented in the school with a rating of either a "2" or "1" (indicating "slightly represented" or "not represented," respectively). In hindsight, it was a poorly designed question because it just said "group culture" without any clarification. Understanding that they were emerging bilingual and multi-lingual learners in my class, I should have framed the wording of "group" to be more specific. On the statement about whether students felt their group was represented in my class, 20% of students (two fewer students) felt underrepresented in my class. Note that I am combining "1" and "2" on the Likert scale, as "3" was neutral. However, when I asked about whether they felt represented in my class as individuals ("I feel like I am represented in this class"), the lowest value was a "2," which accounted for about 10% of students. The responses suggested that 12 people felt that their groups were underrepresented in the school, 10 felt that way about the class, but five felt that way as individuals in the class. It was interesting to me to see the difference in results in what I now perceive as two very similar statements about my classroom. The idea of whether the individual is represented compared to the group culture is an idea that I struggle with still. But for me, these responses support my belief that it is more important to see the individual person and to not pigeonhole them into groups of my design.

I administered that survey because I started to realize that the only thing I really had control over was my own classroom. I could not do anything about the overarching school culture, especially in a larger school; but I could work on my own classroom and hope that some of the positive things that I do bleed into the overall school culture. To me, that was incredibly freeing. I didn't have to worry so much about the school itself because the classroom was the only thing within my control. And as I thought about it, the thing that felt most lacking in the classroom was the inclusion of the students. There was a huge wealth of knowledge—an aspect of students' assets—that was not tapped into.

I made it a point to try to learn more about and use students' cultural assets in my instruction. Trying to delve into their pre-defined cultural groups became increasingly difficult. There were too many students with too many cultural nuances with too many varied interests. Looking at just one class of students, I might have had Peruvian students mixed in with Guatemalan, Chinese, Korean, and Yemeni. But I quickly grew to realize that my students' assets were not necessarily what is traditionally considered "cultural." They were into video games. We spoke about webtoons and basketball. Their interests made me want to know more about them as individuals, not just what was considered a more traditional cultural grouping. Understanding them individually became a larger focus for me as opposed to their grouped assets because it meant that I could actually learn something about the person, instead of making assumptions based on a group of my design. For example, it is easy to pretend that all Chinese students in a class are similar. But there are regional differences, home structure differences, and differences in interests. As the New York State Education Department (NYSED, 2019) phrases it, culture comprises

> the multiple components of one's identity, including but not limited to: race, economic background, gender, language, sexual orientation, nationality, religion, and ability. Culture far transcends practices such as cuisines, art, music, and celebrations to also include ways of thinking, values, and forms of expression. These ways and forms are in constant flux, renegotiation, and evolution. … From this perspective, learning is rooted in the lives and experiences of people and cultivated through activities that people find meaningful. When teaching is not rooted in students' lives, student learning suffers. (p. 11)

So now I define culture as being about who you are as an individual, the perspectives that you bring, and the way that influences how you understand the world. And in hindsight, I think that the discrepancy in the survey results was largely due to a difference in understanding and interpretation of the term "culture" between myself and the students.

Teaching English Language Learners
As a new teacher at The Melting Pot, unfamiliar with teaching a large population of English language learners (ELLs), I worked with a mentor teacher, an instructional specialist, who talked with me about what we could

do in the classroom to make sure students can succeed. We talked a lot about frontloading vocabulary and including various multimedia. We also talked about giving students chances to speak and work in both their home language as well as in English.

It was actually this school that prompted me to join our CRSE research group. I was in an unfamiliar situation and needed ways to make material accessible, and ideas for how to connect to the cultural assets and lived experiences of my students. It was here that I tried various culturally responsive strategies like a protocol to promote student talk and encouraging students to complete assignments in their native language as well as in English, with an opportunity to earn extra credit (Wallace et al., 2022). The Idea Exchange, a protocol to encourage student talk, was modified from the Rumors Protocol developed by New Visions for Public Schools (n.d.). Starting with an open-ended question, with the explicit mention that there were no wrong answers, students were prompted to talk about their ideas with students outside of their usual bubble, typically based on language. In this multilingual and multicultural school, students often sat with those with whom they shared a common language, which is beneficial if they are trying to process information. However, in practicing the multiple modalities of language, it was also important to have them write, speak and listen in English. And it was here that I added modifications to the Idea Exchange protocol, including that new students could listen and write down whatever they could understand in class as a way to practice some parts of the language. The protocol became a way for students to share their understanding of the world, so I could integrate their assets into the curriculum and address any misconceptions, and they could develop their science ideas by sharing with others. For example, we learned about how various countries managed hazard preparedness through students sharing about their own experiences. I moved this success into a larger-scale project, where students explained how people prepare for hazards like earthquakes and hurricanes, citing specific examples of how various places dealt with each hazard (Wallace et al., 2022).

There was one really high note at this school. I taught the first half of the curriculum and then students switched to a different teacher for the second half. This caused me frustration as it seemed that I was just getting to know my students and then they would leave. But once, when my students were shifted to the other Earth science teacher, a student sent me an email along the lines of: "Thank you for working with me. I hope you keep teaching, for me you were the best." Those last words were really what made me feel like I did something of value in that school. I remember the student well, Frances Limo. His English was understandable and he worked well with the other

Portuguese students and helped them with their English. I often paired him with Ana-Sofia, who also spoke Portuguese but struggled with English. I encouraged them to work in their language groups as I was struggling to work with my students' many different languages as a novice teacher. What did it say that my approach was effective for Frances? I like to think that it was the culture that I built in our classroom: That it was okay to be wrong and to make mistakes as long as you take steps to fix it. The next semester, I found out that Frances was failing and was both surprised and a bit angry.

I don't think I did well at The Melting Pot High School. Part of this was due to my lack of experience in working with English language learners. My mentor insisted that what I was doing—the supports and the style and the pacing—helped, but I don't know if I believed her. If I did have any success there, I think it was because I allowed a space for people to struggle and improve. I wanted students to believe that it's okay as long as you are willing to show me your effort and we could work to make sure you do well. This has become a recurring theme for me, and one that I stand by as I continue my teaching career.

Kind Heart Academy for Growth and Excellence: A Lesson That We Should "Love the Kids" (Pre-pandemic)

I remember my interview with the principal at this school after I applied for the teaching position. My past schools were centered on the idea of student success, but in this interview, the principal said that he had one major request for me: "Love the kids." Of all the things that he could have said, that was his message. It was embodied in the culture of the school, what we spoke about and how we operated as a school. It was a beautiful message; it was simple and above all it was kind. The students were people before they were students. Love the kids.

As with The Family Academy for Data Sciences, Kind Heart Academy for Growth and Excellence was a smaller school in terms of number of students and had a student advisory system. I had about eight students in my advisory. There was no overseeing advisory board so each advisor was expected to work with the students in their individual advisory. There were no assemblies for the students to sit with their advisory, no silly names or inter-advisory competitions. It was a space for students to be and to talk about their lives with their advisory, provided they were willing to share. The most positive thing was that it provided a point person for the other teachers to contact regarding small groups of students. I spent a lot of time communicating back and forth with a biology teacher because one of my advisory students was on a constant cycle of "almost failing because of missing work" to "failing because of missing work" to "making up everything for two weeks." It helped to have someone teachers could talk to about it to hold students accountable. There were several instances when, because I knew and closely followed the progress of my eight students, I was able to send a student to their teacher to make up work. It was not the most efficient system, but there were some successes. For instance, I had a student who caught up on multiple lab assignments during advisory time and another who was able to get extra tutoring time. For me, because it allowed the students second chances, it illustrated high expectations for students' academic learning (Gay, 2010; Ladson-Billings, 1995a). You are allowed to make mistakes. It is, however, important to work on doing better.

My first unit in my Earth science class is maps; one of the more challenging aspects of maps is topography. Kind Heart Academy for Growth and Excellence was interesting for this topic because the vast majority of students did not live near the school; rather, students came from various parts of Brooklyn, Upper Manhattan, and the Bronx. For our topography lesson, we talked about elevation and sea level rise. We used an interactive map where we could play around with the changes in sea level to talk about places that would be more susceptible to sea level rise and prone to flooding. (Incidentally, the school itself would have been among the first places to flood due to lower elevation so their greatest commonality would have been destroyed.)

Furthermore, this school was responsive to the needs of its students. For example, there was a point at which the staff said to the administration, "Hey, the students need a mental break. They've been at it for a while and some of them are a bit burnt out." A week later, during faculty time, we were told to stop calling parents and to take that time to plan something fun for the students. We planned movies, games, and clubs that teachers felt were appropriate for a mental break. I teamed with another teacher to host a Super Smash Brothers mini-tournament where kids played against each other and us. The school cared about the students in the purest and kindest sense.

The school culture, largely set by administration and senior teachers, gave students multiple opportunities and experiences. For example, I asked the principal if I could take a few students to the American Museum of Natural History for a unique experience through which students could collect fossils and work with students from other schools as part of the Earth-sciences Reciprocal Learning Year (EaRLY) Initiative (Trowbridge et al., 2020). This Museum was where I earned my teaching degree and I was participating in EaRLY, a science research experience for graduates and their students. The unique experience was to take place over the weekend and would need approval from the school but would not interfere with classes. I chose students with varying measures of academic success, and the principal was happy for me to offer these experiences to students who were struggling academically. In late November, when the event happened, the principal inquired about it and about the students' experiences. I had not mentioned it to him since the initial meeting and he remembered the event and the students by name. This was different from my previous school, because this principal saw students as people, not just grades or numbers. He showed that he cared about these people by remembering and asking me about this weekend activity.

At this school, the students could go outside for lunch; there was a formal lunchroom, but they were not required to sit there. The expectation was that they would come back in time to go to their next class. They often came back late because the line in McDonald's was too long, or everyone went to Chick-fil-A, or the frostee machine in Wendy's was broken so they had to double back to McDonald's. But also in this school a few students regularly stayed behind in my classroom during lunch. We'd just talk about life. Sometimes they'd play UNO or Ultimate Werewolf and invite me and we'd laugh and joke around. It was a space for kids to go, to talk about whatever they wanted or needed. More often than not, I'd be in the background listening and chatting with them.

When the school was struggling, whether with pass rates or attendance or decreased enrollment, teachers were expected to do what we could to help the students. That much is normal. But even in those instances of struggle, the message was not "How can we increase enrollment? How can we improve test grades?" The message was "How do we help the students?" And from the perspective of the principal, I think it was simple. Love the kids.

Kind Heart Academy for Growth and Excellence: A Lesson That We Should Center Students' Well-Being (Pandemic)

Schools closed down in NYC in mid-March of 2020 because of the COVID-19 pandemic. (See Chapter 10.) The last Earth science lesson I taught in person before shutdown was basically an FAQ about COVID-19. I covered what we knew at the time, and what we might expect. I remember ending the class by saying that I didn't think that schools would be closed because of the many social services they provide. I was soon proven wrong.

In the early days of the shutdown, there was a lot of running around and confusion. There were concerns about things like standardized testing, passing rates, and what we should do next. I tried to ground my own work in what I felt the students needed, continuing with the idea of taking care of the students. I started by giving students a break: Three weeks of no new content but with some review. Something new, I reasoned, would have been information overload. They were dealing with life. When they were home, my students had other non-school responsibilities and things they had to deal with. The principal, when I spoke to him, was encouraging of my approach.

A few weeks into remote teaching, I started to run some synchronous classes with asynchronous options. The school, as a whole, had not yet adopted a policy for how remote classes should be run. But I wanted to provide a sense of normalcy for those who needed it. My synchronous class was run very similarly to my usual class but toned down. I ran the synchronous class three times a week with a session time that was voted on by the students in each class to be responsive to what they wanted. I recorded my classes and made interactive video lessons with formative assessment questions as the video progressed. If they were in class and participating, they could ignore the video and the associated assignment; but if they missed class, they had a chance to do the work whenever they could. And there were days when we just spent time with each other, talking about the struggles of pandemic life.

Working in this situation really gave me the chance to think about what my kids needed. My approach was well received in the school and eventually we made it a schoolwide system: Limit the number of days of work, make sure things are accessible, give students a chance to breathe, and check in on their well-being. The school administration was receptive to the idea and it became the premise of our work in the second half of that school year. We made a spreadsheet to make sure we never overloaded kids with too many classes in a day. On any given day, students had two or three classes at most. Basically, we did what we could to make sure the kids were okay.

And there were times when they were not okay. Several had family COVID-19 scares. Some were sick themselves. Many were struggling with the lack of schedule and social interaction. We did our best to make sure that they had a space to talk with each other, to just be kids and to share about themselves. For those who showed up to the synchronous sessions, I think it worked. I had students who felt comfortable to tell me that they were having issues at home and asked if I could forward that information to their guidance counselors. I had students who spent hours talking to me about their struggles. And although I couldn't really do anything to help, I think they needed that space.

When the pandemic started, I had spent only half a year with the students at Kind Heart Academy, having many conversations about their lives. When the school went fully remote, some students voluntarily joined my synchronous class. They wanted to talk both about the science content and about their personal difficulties. I had my students vote on the time of the synchronous class and then one month later they told me that they could not bring themselves to go to their classes because it was too difficult to wake up or they had siblings who are too loud, which made it difficult to focus. I remember several of them messaging me that they were quarantined and afraid to go outside, or that they missed my class and the other students. I remember Kody messaging me that her mom passed away and she did not know what to do, and telling her that she should make sure she spends time taking care of herself and that I would tell her other teachers so she didn't have to worry about it. I remember when Clara, one of the students who regularly stayed in my classroom during lunch, had to quarantine with her mom and how she was so scared for the future. I remember Sheamus telling me about how his mom was in the hospital for a month and how happy and relieved he was when she finally got out.

Lessons Learned

Although I am no longer in the schools I have discussed so far, I carry parts of each with me as I continue my professional journey. At my first school, I learned the value of uplifting students' cultures and that students need opportunities and chances to show that they can do well despite struggling. In my second school, I learned that the only thing I can control is my own space and that, although grades matter, students cannot just be a grade. Pigeonholing students and making assumptions about them is detrimental to their success. In my third school, taking care of the students, being kind and understanding of life situations left one of the biggest impacts on my life. And I like to think that parts of my classroom atmosphere are bound in the best aspects of each of these schools. Some of this is similar to what is often called "critical caring" (Howard, 2021; Williams, 2018).

Bubble Havens High School: Finding My Niche

My current school, Bubble Havens High School, is the most traditional and the oldest of the four schools. It's the type of school that one would think of if they wanted a large high school in a metropolitan area with a population with high percentages of students identifying as Asian and Latine (NYC DOE, n.d.). The school does everything by the book. There are well over 200 teachers. There are assistant principals for each department, there are people who handle things like purchases, there are people who handle staff attendance, there are secretaries for each of those people, and there is bureaucracy as far as the eye can see. But it is within this space that I feel like I truly found my niche.

Even this past year in Bubble Havens High School, a more rigid school that focused on the expectations more than the students, I spent a few minutes each day during the "Do Now" or at the end of the day casually talking with my students. I believe that my students feel that they can talk to me about their difficulties and their barriers to learning. One particular student comes to mind. Our first casual conversation was about her trip to Nashville in the middle of the semester and how she was staying with her sister. I shared about some of the places that I enjoyed there and assured her that she can make up the work and it would not negatively affect her grade. During the semester, life happened and things like her pet passing away and her family going through divorce were conversations that we had while she was working. In the end, we figured things out and she was successful in the course despite life.

I am a teacher in this giant school of over 3,000 students who sees each student as a person. I listen to the small problems, the big problems, movie suggestions, weekend plans, struggles, and happiest little moments. I am by no means perfect but I do my best to care about the students, all of which is grounded in three basic ideas about my classroom related to CRSE.

Three Basic Features about My Classroom

If you walk into my science classroom, you'll see people talking. You'll see people moving around. And you'll see people working on something. You'll see me walking around the room talking to kids, sometimes about work or whatever it is they're talking about. You'll see me raise an eyebrow at something silly they said and them catching that glance and going back to work. I describe my classroom as a space for people to talk with the underlying understanding that things have to get done and they have to get done well.

I ground my classroom in three basic features that connect to CRSE in distinct ways, through: 1) conversation, 2) honesty and trust building, and 3) responsibility and opportunity. Next, I delve into each of these main ideas and what they look like in my classroom.

Conversation: Building a CRSE Classroom through Teacher-Student Relationships

I propose the idea that a casual conversation humanizes a person. CRSE insists on the importance of the cultures and identities of all students. We cannot truly leverage students' assets if we know nothing about them (Mehta & Aguilera, 2020). Building close relationships and partnerships, and learning about students, all begin with conversations.

A conversation with a student can be remarkably powerful. Research has shown the importance of building a positive teacher–student relationship (TSR). The general idea is that a positive TSR will lead to better measurable results in terms of truancy, motivation, and success (Knight-Manuel & Marciano, 2018; Yunus et al., 2011). In some schools that I've been in, the bottom line mattered more than the person and this influenced the way school personnel interacted with students. I will not pretend that the grade doesn't matter. School pass-rates, after all, are tied to school "success." But, in my class, you are not just a student. You're a person. Who you are, how you feel, the things that make you an individual, matter to me.

The basic starting point for this is checking in on students. When in person, I do that by walking around and asking them questions. Online, this was harder. Not seeing faces or reading expressions made it difficult. Periodically, what I did during remote teaching was a quick check-in, and sometimes I wouldn't get a response, which is fine but I spoke to and followed up with those who did. I also gave surveys. For instance, results from a survey question I asked my students offers an example of how I check on their well-being (see Figure 6.1). On that occasion, I learned that two-thirds of my students were feeling normal (66.4%), and nearly half were tired (47%). I also learned that, at that time, none were fearful (0%), and there were equal amounts who were confused (7.4%) as sad or unhappy (7.4%).

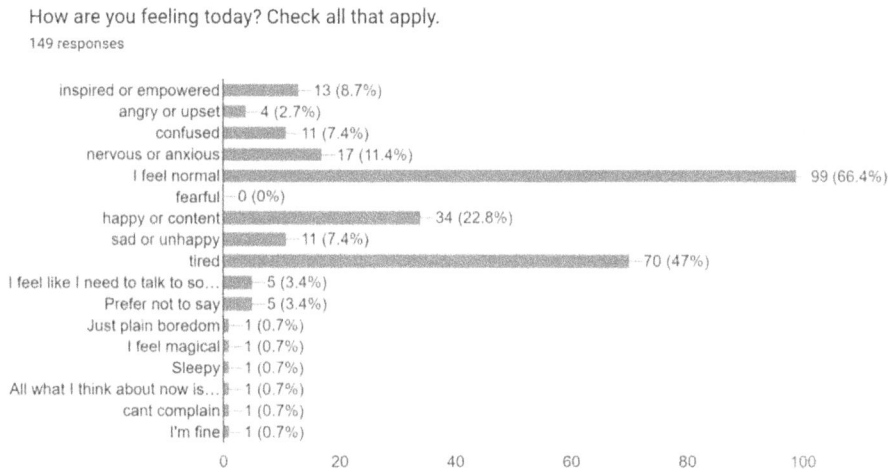

How are you feeling today? Check all that apply.
149 responses

inspired or empowered	13 (8.7%)
angry or upset	4 (2.7%)
confused	11 (7.4%)
nervous or anxious	17 (11.4%)
I feel normal	99 (66.4%)
fearful	0 (0%)
happy or content	34 (22.8%)
sad or unhappy	11 (7.4%)
tired	70 (47%)
I feel like I need to talk to so...	5 (3.4%)
Prefer not to say	5 (3.4%)
Just plain boredom	1 (0.7%)
I feel magical	1 (0.7%)
Sleepy	1 (0.7%)
All what I think about now is...	1 (0.7%)
cant complain	1 (0.7%)
I'm fine	1 (0.7%)

Figure 6.1 Results from a student survey question indicating how my students were feeling at the time. Graph developed by the author.

Connecting to Students' Lives

In my third year of teaching, I started to go to NYC Comic Con (an event geared towards the more nerdy parts of U.S. culture, with superheroes, cartoons, comic books, and more). One reason I went was to get some comic books for students to read. I reasoned that an interest in reading can be sparked not just by literature but by visual novels as well. I purchased around 30 comic books, left them in my room, and told my students that they could take them if they wanted. Although most were left untouched, it started to lead to conversations about things like webtoons, a Korean-style of online comic books, manga, anime, cartoons, TV shows, and whatever else came to mind. Pretty soon, my students started talking to me about their reading habits and things that they enjoyed.

Using things students already know is a basic part of science teaching. It means going back and connecting to students' lives, drawing upon their assets and cultures, validating and building up their connections to their cultures. A prime example of this is an assignment that I gave using local geology where students had to take photos related to content we had learned in class: rocks, weathering, erosion, landscapes, plate tectonics. The mock presentation that I provided as a model included things like the rusted school gates, the worn-down steps of the school, and the cracks on the sidewalks caused by old tree roots. I wanted students to note that everything that we were learning was about things that they see everyday, but now they had the vocabulary to identify it and the understanding to describe what was happening.

Local Geology Assignment: Instructions for Students

1   Take five photos related to the geology content we are learning in class. This could be rocks, weathering, erosion, or another aspect of science you can see in your local surroundings.
2   Include yourself or some form of identification (for example, your school ID) in each photo.
3   For each photo, write a description explaining

   • How the photo connects to your life, and
   • How the photo relates to what we're learning in class.

4   Prepare a presentation to share your photos with the class.

I use some of my understanding of my students to support their learning. For instance, I give them the option to start assignments in their home language and translate it into English. As they become more proficient in English, I insist that they try to write first and then translate back to their home language, to see if it makes sense. It can even be extra credit, as was the case when I asked them to explain how we protect ourselves from natural hazards by creating safety pamphlets and include how their home countries prepare and their personal experiences with these phenomena (Wallace et al., 2022). I will absolutely use the rusted school gates as a conversation piece about chemical weathering. We can talk about why the docks of some Caribbean Islands have more rust than the docks near our school, a comment made by my Jamaican and Haitian students. We can share why air conditioners are placed closer to the ceiling and why heaters are placed on the ground. We can use anime to talk about rock formation and landscapes, and we can use home countries to talk about how places have historically prepared for certain natural hazards. But none of that happens without knowing more about my students.

In my experience, making conversation central to the classroom environment has led to an increase in student engagement. For instance, one conversation we had was about the webtoon "Unordinary," which, at the time of this writing, was about a teenager's struggle with his mental health as he searched for a way to fit in. While casually discussing the last chapter of "Unordinary," I said, "Okay but we also have to get through this class, so let's turn back to the work and let me know if you need help." And the students went to work. In having this offhand conversation, I showed the students that I saw them and then redirected. It becomes much easier to redirect a student when your relationship is not just authoritarian or driven by fear (Walker, 2009). The positive interaction led to greater buy-in and engagement from the students (Walker, 2009; Yunus et al., 2011).

Even during remote teaching, this was something that I tried with various degrees of success. My classes often started and ended with conversations about current events, what was happening in their lives, or anything else. It gave them a space to be. Resoundingly, my students said that they appreciated that mental break and were able to spend some time just to exist.

In consideration of what the students needed, an idea that I took from my third school, I administered multiple surveys throughout the year to check in on student progress and see if anything needed to change for the class to flow better for them.

Questions you might ask your students for feedback

As a way to understand what I should change moving into the next year, I asked my students questions in my final survey of the year. I've adapted the questions that I posed to my students to ones that you might consider asking yours:

1 If you were to talk to yourself at the beginning of the year, how would you describe our class? Think about the structure and culture in our classroom.
2 Please explain your answer to the question: "Did you enjoy the class?"
3 What is one thing that you would change about this class? Try to be specific and include how you would do that.
4 Think about yourself in the class. What did you do? Were you a positive part of the classroom? Did you struggle? For example, if you felt like you were consistently working and were comfortable talking in class, that's something that you can mention. Or if you felt that you did not feel comfortable and struggled, please describe that.
5 How about the teacher in the class? What did they do? What did they contribute to the class that you liked or disliked?

For my final survey of the year, responses suggested that, overall, students appreciated the calm and caring atmosphere of the class (see Figure 6.2). There were comments citing the mental break minutes, the random conversations, and the joking atmosphere. These are things that were largely in my control, so it was really about what I could do. Representative student comments include:

- *I was definitely always working and very comfortable telling Mr. Tsoi anything and not being afraid of getting in trouble or that he didn't trust me. Which is also a main reason why I love coming to this class.*
- *He contributed a sense of something to relate to because he would talk about things that were relevant to me. He would talk about random video games or random movies/shows and some random piece of news and that was something that I liked.*
- *I definitely enjoyed attending class because Mr. Tsoi is an understandable teacher. He always made us feel comfortable in the class. He would joke around, let us talk if we needed to and he would try to make time to help if there was a question, unlike in other classes [where] I don't feel comfortable asking for help because I felt judged.*

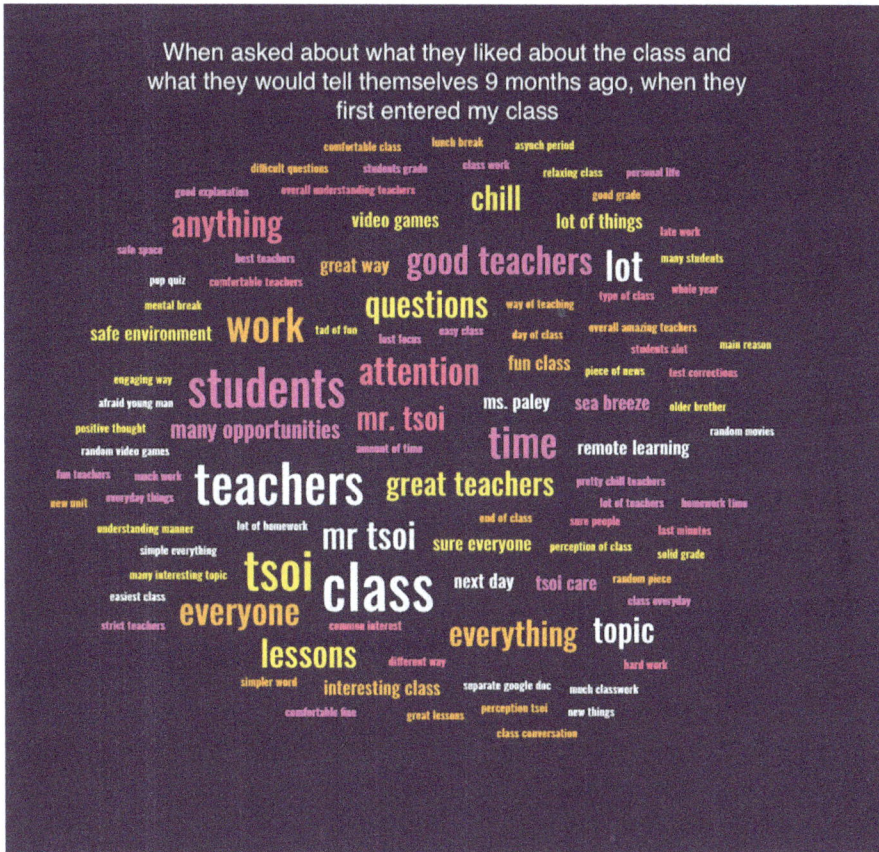

When asked about what they liked about the class and what they would tell themselves 9 months ago, when they first entered my class

Figure 6.2 Word cloud developed from student survey responses about what they would tell themselves about the class if it were just starting. Image courtesy of Tsoi, 2022.

The more critical comments were around the struggles with school in general. One person noted that they were "Not motivated to do school," but there would be others including the chronically absent, who likely felt the same way. After all, the data here was pulled from students during remote teaching, so those who chose to come to class wanted to or had the capacity to be there. Another student noted, "This year I've been busy," which was a reality for a lot of my students. As mentioned earlier, while at home, students had other responsibilities, such as taking care of family or working. In this particular survey, while I had 55 responses that were almost entirely positive, ideally I should have had closer to 150 responses total since I had five classes each with about 30 students.

Honesty and Building Trust

Relationship building in CRSE involves trust. You must be honest to know the students. If there is a loss of trust then there becomes a loss of respect and that TSR becomes fractured. A successful classroom requires student buy-in. Without honesty, you cannot build the trust necessary to have that type of classroom. As noted throughout this book, relationship building between teachers and students is an important part of learning about the students, and is supportive of developing other aspects of CRSE (NYSED, 2019).

Throughout these conversations and in any interaction with students, I make sure I'm honest. Honest about my screw-ups, about their progress in class, about my intentions and goals in the things that I ask them to do. Always just be honest. There were days when I went into class with a sour mood. Is that unprofessional? Probably a little bit, but like everyone, I was human and I had some good days and some really bad days. And sometimes things add up and you just don't have that motivation. And I shared that with them. It was often simply: "Hey, I'm just not having the best day and I would appreciate it if you could x, y, z," or "Full disclosure, I had a terrible morning. I may not be up for the usual banter." Something that showed them I was human and willing to be honest with them. Unsurprisingly, when I had those days, the students were very willing to pick up the slack. The following day, I would often apologize for my mood and we would move on (Hosek & Thompson, 2009).

I did not sugarcoat things. When a student was not doing well in class, I pulled them aside (or in a breakout room when we were remote) and told them, "This is not a representation of your skill. This is terrible. I expect better because I know you can do better." I've told them that there were points when something was not worth the effort of doing, that it was okay to take that failure and do better next time. And that struggles are a normal part of life. I've spoken to them about my own struggles and failures, what I've done to improve, and am honest about the mixed successes of that. It's important for them to know that I am not perfect, that I too struggle and make mistakes, and am trying to do my best, and expect that from them.

I had a student at Kind Heart Academy for Growth and Excellence (before the pandemic) named Grady. Grady was always engaged in class, asked a ton of questions, joked around, and generally performed well. It was around January when she did not perform to my usual expectations of her and I made an offhand comment as she was leaving for lunch: "Not nearly close to what I expected of you." She stayed in my classroom during her lunch period, and we were joined by a group of students who often came to my classroom for lunch. Even though she didn't know these students well, Grady opened up about some of her struggles, talking about mental health, the pressure of not reaching other people's expectations, and responsibilities at home. The lunch group listened and encouraged her and provided a space for her, sharing their own struggles—the good, the bad, everything in between. In addition to the character of the students, I like to think that some of this was due to the space I provided and encouraged.

Responsibility and Opportunity

Related to the CRSE tenet of holding high expectations, responsibility is needed to demonstrate success, improvement, and giving students ownership of their education. The best way to show ownership is by having students take agency of their own work and of their grades. Giving students chances to improve is a way to show that you care, to offer opportunities to bring themselves to passing, but the responsibility for learning is largely on them.

Responsibility and opportunity, for me, are often two parts of the same problem in school. Students need to be allowed to fail. For all of the talk about "ownership of their work" and "high expectations," some schools have fallen into a trap where numbers are fudged to pass more students (Howse, 2019; Shaban et al., 2019) because pass rates are considered an indicator of success. For example, in NY, one metric used to determine school quality is graduation rate. Also in NY, there is a "college readiness" score which is based on the state exams for Algebra 1 and English. A student who receives higher than a 75 in English and an 80 on the Algebra 1 state exam is considered "college ready" (Chen, 2023). So teachers "teach to the test" for numbers to look good, and everyone goes on their merry way. Doing so allows the students to get a better grade, makes the school look better, and teachers look like they're doing a good job. The first school I worked in was proud to say that 100% of their students were accepted into college and mentioned it during their first graduation. There was of course no mention of how teachers regularly stayed after hours to allow kids to make up work, or how many kids would go on to fail their first year of college classes. One of my most capable students from that year emailed me that he failed his first year of college. But yes, there was technically a 100% college acceptance rate.

In a more perfect world, grade inflation would not happen because it wouldn't be a reflection on the school. In order to get students to buy into their work, they have to be responsible for their work. Why should a student care if they will be passed upwards regardless of submitting their work?

How does an inflated grade reflect their work? Why is "good enough" considered good? We cannot suggest ownership if students are not responsible for their work. And students should be allowed to fail but have the opportunity to show subsequent success. As I say often in class, "It's okay to fail as long as failure is not your last step." Screw up. Fix it. Do better next time. It is also here that the classroom environment is most aligned with CRSE and the idea of high expectations. High expectations should be the premise of school, especially if we're going to truly educate students. It is for this reason that I try to balance my gradebook to show successes after failures. I have implemented "fix it days" for classes that are really struggling to give them an opportunity to improve and succeed, but never to cruise into a passing grade.

Student Survey

In a mid-semester survey that I gave in Bubble Havens High School to check on the progress so I could adjust pacing and expectations, my students resoundingly responded that they appreciated the support and the opportunity to do better. The results suggested that under 5% of students felt that they were not responsible for their grades and less than 4% felt that they were not given opportunities when asked, "Any final comments about the grading of the class? It can be good or bad, or if you feel like you need to justify your answer to something."

Illustrating several students' ideas from the mid-semester survey, one student responded, "I stated earlier on in the survey how Mr. Tsoi is very caring and always wants the students to do well. He gives plenty of opportunities for students to make up work, do extra credit, or basically anything to boost up our grades." Another wrote, "The grading of the class is good because they tell you why you got it wrong and explain how to fix it." These responses encouraged me to continue with my grading policy. Additionally, I believe that these survey results told me that there was some work where students could see the results of their effort. There were two negative responses indicating that these students believed the exams were hard or they wished there were more chances to review. As one wrote, "If I were to add something to the class it would be before doing an exam we would do a review on every topic that would be on that exam and we would be given homework on those topics. I feel like it helps a lot with memorizing and understanding the topics."

Responsibility and opportunity is not about the students alone. In any relationship, there has to be effort by both sides for success. So, what is our responsibility as teachers? What is it that I contribute that makes the classroom work for students? I tried to ask students this question but the most common responses have been some variation of "just make sure you teach me things." What that told me was that the question itself was poorly written. The version of the question that I ended with was: "What did Tsoi contribute to the class that you liked or disliked?" A lot of the responses that were already shared were from that question. Based on what I've learned from my students' survey responses and in my many spaces over a short period of time, I think my biggest responsibility as a teacher is to provide a comfortable space for the students to learn.

Moving Forward

As the school reopened for the new school year (2021–22), and we got back into the groove of physically being in school, I had a few concerns. "Learning loss" was one of them, as it was the only thing being broadcasted loudly through news outlets (Calvert, 2022; Mervosh, 2022). My biggest concern was the mental state of the students. I am glad to say that a lot of students came back to school hesitant, but ready to re-engage in the community.

Students stop by to say "Hi" and chat a bit. Some I've taught in person, some from the hopefully bygone era of remote learning. We spend time talking about the previous day, what they're concerned about, shows and games, talking and being human. It's funny, as I'm writing this, I realize I never really thought of myself as that person, the person to be there as an adult listening to the students and their stories, their struggles, their humanity.

My students generally appreciate, I think, the safe space, the space where it's okay to mess up and you won't be judged and you are free to struggle, academically or otherwise (Holley & Steiner, 2005). During the second week of school, I had a student who proclaimed that she never liked her science classes because she doesn't get it. She also regularly went to the bathroom for a walking break and was on her phone when in class. My challenge to her premise was if she didn't "get it" because she was on her phone or if she was on her phone because she didn't "get it"? From then, I started checking in on her more often, at least twice per period. The bathroom breaks slowed. Her work improved and she got into the habit of telling me what she did understand before asking more specific questions. All of these are important parts of her growth. And all of that was creating that classroom culture.

In the remote space, the common characteristics (mentioned in a student survey) were that my class was calm, understanding of the students, and there were opportunities and that you should not be afraid of getting anything wrong. The first part of setting up the environment is based on the trauma-informed practice of "being the thermostat," where you set the tone of the room (Hagenauer et al., 2015). Not too hyper, not too low, but calm is what I hope for and for many of my students it showed. The idea of being understanding of the student is based partly in CRSE, but also based on listening in on their conversations. They are, as mentioned before, people in the room and not just grades on a spreadsheet.

I share a summary of responses from a student survey administered recently. It is important to note that the comments here are strictly positive. During pandemic teaching, in a fully remote space, there were many students who did not show up. Those students would likely have had negative responses or would have chosen to not do the survey. The ones who did show up, seemed to appreciate my class. For instance, two commented, "it was kinda easy and because I also felt welcomed" and "lessons were always very in-depth and I never felt left behind. If I did, Mr. Tsoi would always give extra help and explanation." Another noted, "[the teacher is] calm and understanding of whatever situation you're in. He has a different way of teaching than most teachers. and I think that is the main reason why I know many students are passing his classes like me."

Opportunities also means that messing up is an option and a way to grow. Being stuck in the idea that everything must always be nice and perfect is unhealthy for the students and for the teachers. So it becomes important to note that the classroom cannot always be the most comfortable space. Learning is uncomfortable in the sense that you are approached with something you do not know or understand. Safety and comfort have to be balanced (Holley & Steiner, 2005). Chances to be wrong. Chances to improve. That is opportunity.

Conclusion

It is my hope that the work that is done in the classroom will slowly bleed into the culture of the school, which is much harder to work on, especially for me since I have popped around between so many schools. And some things that I do will not mesh well with a particular school. A more rigid school, one that is regimented and structured, would not fit me or my work because a lot of those types of schools have a tendency towards working on acceptable perceived success (also known as test scores).

But my work, and the things that I care more about, is geared towards centering the students in the classroom, in a way that is genuine and true to myself. There is a lot of banter, a lot of joy, an understanding of the human behind the pens and papers, or in the Zoom times, the black screen. We laugh, we struggle, we live. People.

Currently I am working in the CRSE community of my new school. It is a group of about seven teachers who are dedicated to making sure the school is more responsive to the needs of the community. We have since branched off into two groups, one focusing on curriculum and the other on school events. I have been listening to both sides. I've been asked recently to choose one side and, to be honest, I don't know which fits me more. But I do know that it's something I want to continue: learning about and helping the students in a way that is authentic to me. It's ongoing work, understanding the populations that we have and trying different things to engage the community.

There can be no checklist for CRSE because the work itself means that you work with the students and who they are as people. So the idea of formalizing is both daunting and somewhat insane to me. How can I tell you how to talk to your students in a way that's real to you? My best work, I would argue, is still in my own classroom space where, according to administration and my students, I have managed a "big brother teaches you things" type of vibe. But the CRSE group at school is working on incorporating the type of openness that is within my classroom into the larger picture by outwardly and explicitly showing more support towards traditionally marginalized groups. In the 2020–2021 school year, we ran teacher professional development on responsive education and how that would support traditionally marginalized groups, some of it as a reminder that this is what good teaching looks like (Ladson-Billings, 1995a). We largely focused on an inclusive curriculum. For the science department, we mentioned finding scientists of color and locations outside of NY that may be more representative of our student population. We have successfully run a teacher professional development about inclusive practices and how the school community can be more proactive in incorporating and supporting the LGBTQIA+ community. And we have worked towards improving success rates for language learners, largely focusing on various ways to support them using language cognates, using translation as extra support, and with images to define vocabulary.

What I'm hoping for in the long run is to end up working in a school that cares about the students. Not just test scores but the whole student. And I hope to be an aspect of that change. A school should celebrate the students and appreciate their diversity and experience. That starts in the classroom. That's the grassroots work. And then it can spread outward to the school and, hopefully, further.

Reflection and Discussion Questions

The way that I run my classroom is with joviality and activity because it is fitting for me. It lends itself to having random conversations, which in turn allows me to get to know my students better. So my classroom has a lot of physical space for me to roam around. And for the most part, I don't mind conversation and I let students work in groups and pairs for many assignments.

1   How is your classroom currently structured?
2   What avenues, not structures, do you use to get to know your students?
3   How have you incorporated what you know about your students into your structures?

Students want to do well, so they appreciate structures that allow them to do well. This can look like many things. For me, I want students to have chances to mess up and then do better. For me, that looks like test corrections for half credit with an included notes file that is created for the unit, open-book quizzes, redos on homework or classwork, and extra help after school.

1   What opportunities do you provide for your students to improve?
2   In what ways are these opportunities achievable when considering other aspects of the students' lives?

Notes

1 All school and student names are pseudonyms.
2 In New York State, there are courses that prepare students for a standardized test known as Regents exams. There are Regents exams in all of the major subject areas. The number of exams a student passes determines the type of diploma they will receive upon graduation.

References

Calvert, S. (2022, September 6). Schools are back and confronting devastating learning losses. *The Wall Street Journal.* https://www.wsj.com/articles/schools-are-back-and-confronting-devastating-learning-losses-11662472087

Chen, G. (2023, March 16). *New York City schools: Deficient college readiness.* Public School Review. https://www.publicschoolreview.com/blog/new-york-city-schools-deficient-college-readiness

Gay, G. (2010). *Culturally responsive teaching: Theory, research, and practice* (2nd ed.) Teachers College Press.

Hagenauer, G., Hascher, T., & Volet, S. E. (2015). Teacher emotions in the classroom: associations with students' engagement, classroom discipline and the interpersonal teacher-student relationship. *European Journal of Psychology of Education, 30*, 385–403. 10.1007/s10212-015-0250-0

Holley, L. C., & Steiner, S. (2005). Safe space: Student perspectives on classroom environment. *Journal of Social Work Education, 41*(1), 49–64. 10.5175/jswe.2005.200300343

Hosek, A. M. & Thompson, J. (2009). Communication privacy management and college instruction: Exploring the rules and boundaries that frame instructor private disclosures. *Communication Education, 58*(3), 327–349. 10.1080/03634520902777585

Howard, T. C. (2021). Culturally responsive pedagogy. In J. A. Banks (Ed.), *Transforming multicultural education policy & practice: Expanding educational opportunity* (pp. 137–163). Teachers College Press.

Howse, A. (2019, July 30). *Toronto high schools are allegedly letting students buy higher grades in exchange for cash.* Narcity. https://www.narcity.com/toronto/gta-high-schools-faking-grades-in-exchange-for-money

Knight-Manuel, M. G., & Marciano, J. E. (2018). *Classroom cultures: Equitable schooling for racially diverse youth.* Teachers College Press.

Ladson-Billings, G. (1995a). But that's just good teaching! The case for culturally relevant pedagogy. *Theory Into Practice, 34*(3), 159–165. 10.1080/00405849509543675

Ladson-Billings, G. (1995b). Toward a theory of culturally relevant pedagogy. *American Educational Research Journal, 32*(3), 465–491. 10.3102/00028312032003465

Ladson-Billings, G. (2014). Culturally relevant pedagogy 2.0: aka the remix. *Harvard Educational Review, 84*(1), 74–84. 10.17763/haer.84.1.p2rj131485484751

López, F. A. (2017). Altering the trajectory of the self-fulfilling prophecy: Asset-based pedagogy and classroom dynamics. *Journal of Teacher Education, 68*(2), 193–212. 10.1177/0022487116685751

Mehta, R., & Aguilera, E. (2020). A critical approach to humanizing pedagogies in online teaching and learning. *International Journal of Information and Learning Technology 37*(3), 109–120. 10.1108/ijilt-10-2019-0099

Mervosh, S. (2022, November 28). Pandemic learning loss. *The New York Times.* https://www.nytimes.com/2022/11/28/briefing/pandemic-learning-loss.html

New Visions for Public Schools. (n.d.). *Rumors: A group learning routine.* Retrieved March 21, 2022 from https://docs.google.com/document/d/1EO9FFg674DzAkXdC3XTtzxAHGiQrh-yfyXAbFtvPj6w/edit

New York City Department of Education (NYC DOE). (n.d.). *School quality snapshot 2021–22.* Retrieved July 3, 2023, from https://tools.nycenet.edu/snapshot/2022

New York State Education Department (NYSED). (2019). *Culturally responsive-sustaining education framework.* http://www.nysed.gov/crs/framework

Paris, D., & Alim, H. S. (Eds.). (2017). *Culturally sustaining pedagogies: Teaching and learning for justice in a changing world*. Teachers College Press.

Shaban, B., Bott, M., & Villarreal, M. (2019, September 5). *Oakland educators accused of falsifying transcripts to boost students' grades and graduation rates*. NBC. https://www.nbcbayarea.com/news/local/oakland-high-school-accused-of-falsifying-transcripts-to-boost-students-grades/1959637/

Trowbridge, C., Stokes, P., Schreiber, H. A., & Hopkins, M. (2020, October 26–30). *Bringing teachers into the field in order to bring field-based paleontological research into the classroom* [Poster presentation]. Annual Meeting of the Geological Society of America, online.

Tsoi, K. (2022, April 21–26). *Developing classroom community through developing relationships* [Poster presentation]. American Educational Research Association Annual Meeting, San Diego, CA, United States.

Walker, J. M. T. (2009). Authoritative classroom management: How control and nurturance work together. *Theory Into Practice, 48*(2), 122–129. 10.1080/00405840902776392

Wallace, J., Howes, E. V., Funk, A., Krepski, S., Pincus, M., Sylvester, S., Tsoi, K., Tully, C., Sharif, R., & Swift, S. (2022). Stories that teachers tell: Exploring culturally responsive science teaching. *Education Sciences, 12*(6), 401. 10.3390/educsci12060401

Williams, T. M. (2018). Do no harm: Strategies for culturally relevant caring in middle level classrooms from the community experiences and life histories of Black middle level teachers. *RMLE Online, 41*(6), 1–13. 10.1080/19404476.2018.1460232

Yunus, M. M., Osman, W. S. W., & Ishak, N. M. (2011). Teacher-student relationship factor affecting motivation and academic achievement in ESL classroom. *Procedia-Social and Behavioral Sciences, 15*, 2637–2641. 10.1016/j.sbspro.2011.04.161

7

Building Relationships to Support Relevance

Reflecting on Culturally Responsive and Sustaining Science Teaching

Susan Bullock Sylvester

HIGHLIGHTS

- In this chapter, Susan explores the questions: *How do I learn more about my students' responsibilities and career goals? In what ways do I support them in these goals in my science classroom? How can my learning about culturally responsive and sustaining education support these efforts?*
- Drawing on existing research, Susan uses a 3R framework to explore these questions through the lens of Relationship, Relevance, and Reflection in her practice.
- Susan features multiple concrete classroom strategies, including a "getting to know you" questionnaire for relationship-building and memory maps to incorporate students' voices and experiences.
- We hope you will take away from this chapter one teacher's efforts to grapple with and internalize the CRSE tenets, along with tangible strategies to consider and adapt for your own context.

DOI: 10.4324/9781003397977-10

Introduction

I teach science in a New York City public high school that serves students who are at risk of dropping out of school or have fallen behind in earning enough credits to graduate. My students have inspired me to explore the questions: *How do I learn more about my students' responsibilities and career goals? In what ways do I support them in these goals in my science classroom? How can my learning about culturally responsive and sustaining education support these efforts?* Through the lens of culturally responsive and sustaining education (CRSE), I seek ways to co-create with my students experiences of "doing science," and learning about the diverse community of scientists and engineers that collaborate to solve real-world problems. At the same time, I am working toward what it means to implement CRSE in my science teaching. I use student questionnaires, discussions, and reflective writing to explore these research questions.

Impressions from student teaching, my engineering background, and a desire to understand my students' needs and goals has informed my inquiry into CRSE. For me, a main takeaway from student teaching was the power of building relationships with students. I carried this gem forward with me and found that my teaching improved as I made adjustments to my plans based on knowledge I gained through communicating with my students. With relationships as a foundation, I have been considering a framework to help CRSE become more concrete and tangible in my practice, by centering my teaching around three well-established concepts: relationship, relevance, and reflection. In this chapter, I discuss these concepts supported by examples from my own teaching.

I am not certain where I first narrowed CRSE down to basic teaching concepts that I could process more easily. Just as my students build on prior knowledge, I too build on many pedagogical frameworks coupled with daily experiences. The Rigor/Relevance Framework™ (R/R) (Jones, 2010) is one of many frameworks that I have been exposed to directly and indirectly. The R/R Framework combines elements of Bloom's Taxonomy (Bloom et al., 1956) with Webb's Depth of Knowledge (DOK) (Webb, 2002). Researchers have developed frameworks with variations of these "R"s, such as Zaretta Hammond's (2014) Ready for Rigor Framework which connects to culturally responsive teaching.

My purpose in this chapter is not to claim expertise in CRSE, but to provide insight for anyone struggling with the uncertainty about how to incorporate CRSE into their teaching. For me, CRSE is a means to examine my practice in light of who my students are and what builds their confidence as I help them develop scientific thinking skills that prepare them for many of the challenges they will face in the world.

School Context: Teaching at a Transfer School

My teaching journey began in 2015, after a 25-year career in civil and environmental engineering, with the opportunity to pursue my Master of Arts in Teaching degree and teacher certification. The majority of my seven years of teaching has been at a transfer high school in New York City (NYC) that serves predominantly Latinx and Black students who tend to be older than the typical school-age range in most high schools in this country. Transfer high schools are designed to provide a safety net for the hundreds of students who are not on track to graduate with their high-school peer group (Alliance for Excellent Education, 2011; Good Shepherd Services, 2007; Steinberg & Almeida, 2010). In NYC, students may choose from nearly 60 transfer high schools located across the five boroughs, where they are provided with additional support services. The Learning to Work program that provides opportunities to earn income through paid internships is one example (NYC DOE, n.d.).

The reasons students choose transfer schools are as varied as they are, and the circumstances that brought them to this type of school do not automatically resolve because they have a fresh opportunity to gain a high school diploma through the age of 21.[1] This diverse group of young adults are independent thinkers trying to assert their independence in ways that may have presented missed opportunities for academic success within a traditional four-year high school. Some of the challenges we face in transfer schools are students' high absenteeism rates, the appearance of a lack of motivation toward school, and reluctance to engage academically. For many students, however, the challenges are external to school, and may include caregiving or working to support themselves or their families. Because of these challenges, many of my students have become independent thinkers far beyond their years. My focus is on how to help my students use this independence in a manner that engages them in learning science, as I introduce them to science and engineering career paths, and aim to help them pursue their personal goals.

Reflecting on Learning to Teach

I had the privilege of becoming a science teacher after a career working in water resources engineering. I arrived at my graduate-level teacher residency classroom with many ideas from my work experiences. In my first semester, I was assigned to a seasoned mentor teacher in a self-contained middle-school science class. I remember how impressed I was with the ways my mentor teacher facilitated an interesting and welcoming learning environment. The setup of the small class allowed for easy interaction and engagement between the adults and students. The learning activities were designed to foster curiosity and the students were supported through group discussions. At the time, I did not fully appreciate the significance of 12 students with one teacher and a paraprofessional. This model felt very much like the project teams to which I was accustomed from the workplace.

In my second semester residency, I was assigned to a large, urban, high-need high school. The class setting and the approach to teaching by my high-school mentor teacher was very different from the self-contained middle-school science class with the smaller student-to-teacher ratio. The learning activities were focused on helping students understand material to pass the state's standardized test for Earth science and consisted mainly of slide presentations, worksheets, and lab practice. Some students tended to cluster toward the back of the room, sliding low into their desks when the lesson began. I would often hear voices whenever I started to talk. Attendance was sporadic and some students came to class up to 20 minutes after the lesson had begun. As soon as I put students into groups to work on an activity, many would begin talking to their neighbors about topics that had nothing to do with the science we were doing. A few students had their heads down on the table. When I asked students why they were not discussing science topics, I often heard "Look Miss, I'm doing my work!" This response always nagged at me. I have realized since then that "doing my work" wasn't what I hoped to hear from students. I wanted them to be engaged in learning science. This apparent lack of engagement made me question my decision to teach, and I felt intimidated and dispirited.

Fortunately, while I was at this high school, I had opportunities to visit and talk with fellow residents who taught in similarly large classes. In these classrooms, I observed different ways to organize students into groups, and saw examples of active learning opportunities where students were more in charge of their learning. Seeing such different approaches to teaching the same subject left a deep impression on me, and a hopeful one.

Perhaps because teaching is my second career, I see education through the context of the workplace. In my experience, education was the entry point to a career in science and engineering. During my career I supervised a range of people from entry-level technicians to experienced engineers. I learned from many of the technicians that they had not been able to pursue a college degree, but found opportunities to gain an associate's or bachelor's degree more accessible once they were working. I recall a technician who had been an auto mechanic; he became interested in engineering through that work. He gained confidence by pursuing an associate's degree, which led him to become a civil engineering technician. The experience of being a technician further allowed his progression through a master's degree in civil engineering with corresponding job advancement. The knowledge I gained from getting to know the people with whom I worked strengthened my desire to pursue teaching when I retired from engineering. In my teaching residency, I was about to find out that I also had a lot to learn from my students.

At the time of my residency, I was beginning to hear about CRSE. However, learning about myriad pedagogical frameworks, practices, and curriculum requirements was consuming my novice teacher energy. Even with more than seven years of teaching experience, I panic at being asked by an administrator how I am implementing the principles of the New York State (NYS) Culturally Responsive-Sustaining Education (CR-SE) Framework (NYSED, 2019) in my classroom. I worry that I may not even be able to explain what comprises CRSE in the context of lesson planning and curriculum implementation. Gloria Ladson-Billings shares from her own lived experiences and academic research that "students must experience academic success; students must develop and/or maintain cultural competence; and students must develop a critical consciousness through which they challenge the status quo of the current social order" (1995a, p. 160). Intellectually, I could process these ideals, but I felt challenged by what this really looked like in practical ways within my science classroom.

Integrating CRSE in My Science Classroom: Considering a 3R Framework

Through the day-to-day classroom experiences with my students, professional learning opportunities, and an understanding of the measures for which I am assessed (i.e., The Danielson Framework for Teaching [Danielson Group, 2022]), I regularly reflect on my teaching. In addition, I have been a part of a teacher professional learning group focused on learning about CRSE and how it can be applied in science classrooms. My participation in the Culturally Responsive and Sustaining Education Professional Learning Group (CRSE PLG) has helped me to deepen my understanding of the principles of the NYS CR-SE Framework (NYSED, 2019) and provided me with a collaborative environment in which to examine my teaching. I did not necessarily need to reinvent my teaching, but to become reflective and intentional about how I use what I learn about my students to create more equitable learning opportunities. With relationships as a foundation, I have been working to incorporate the tenets of CRSE into my teaching practice. I realized that I had been centering my teaching around several well-established concepts: relationship,

Relational   Relevant

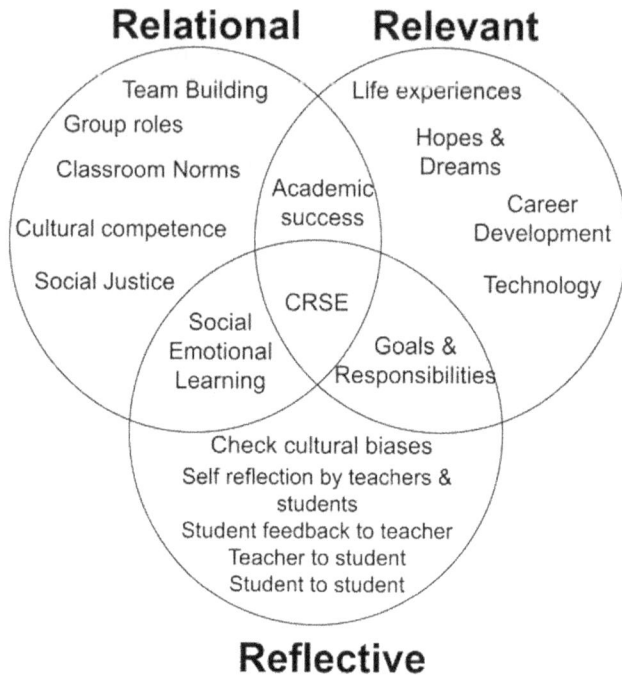

Team Building
Group roles
Classroom Norms
Cultural competence
Social Justice
Academic success
CRSE
Social Emotional Learning
Life experiences
Hopes & Dreams
Career Development
Technology
Goals & Responsibilities
Check cultural biases
Self reflection by teachers & students
Student feedback to teacher
Teacher to student
Student to student

Reflective

Figure 7.1 A visual representation of a 3R framework. Image designed by the author.

relevance, and reflection. (See Figure 7.1.) Next, I examine each of these "R" concepts in my teaching practice and provide examples of how I have implemented them in my Earth science classroom.

Relationships

The first of the three Rs, and maybe the most impactful for both students and teachers, is relationships. I have found that getting to know my students is one of the most interesting parts of my work. Proactively developing ways to build relationships one student at a time is fundamental to teaching and to CRSE. The strategies that I use to get to know my students include student questionnaires, quick check-ins during class, one-on-one conferencing, and career-focused mini-projects.

A "Getting to Know You" Questionnaire

I start the school year with a questionnaire and other "getting to know you" activities. With the need to meet curriculum requirements, I feel pressure to jump right into content after that initial questionnaire and not create ongoing space for really getting to know students. I have to remind myself that a questionnaire is a starting step, as relationships take time to cultivate and require layers of interaction. One way I incorporate questionnaires as a check-in tool and relationship-building effort is by using Google Forms. I use Google Forms because it is easy to keep track of student responses, and to look for connections across all responses. More frequent questionnaires may also help students who are shy give additional insight into themselves. Interestingly, students seem to appreciate the questionnaires, and I give credit to those who write substantive responses.

Some of my students write "idk" (meaning "I don't know") in open-ended responses. When a student responds with "idk," they have actually begun to open opportunities for dialog. For example, in a one-on-one situation, or through written feedback, I may write, "Thank you for acknowledging you don't know. I look forward to hearing more from you about what you mean by 'idk.'" It may be that a student is not sure how to write their thoughts but is willing to talk about them in a different setting, thus allowing me to differentiate in specific ways. All student responses can help to build relational insights; although some initial responses may take longer to understand, I find ways to deepen communication with each student.

My "getting to know you" questionnaires have changed both in approach and in the types of questions that I ask. In my early years of teaching, I was focused on finding out what motivates students in school and would ask questions like "What is your favorite subject and why?" and "What teachers helped you feel the most motivated and how did they do this?" Over time, I have begun to think that my focus on motivation is not productive. As I have built relationships, I have learned that many of my students are balancing a range of responsibilities, such as caregiving and working to support family members. With this insight, I am now asking more about what responsibilities students have outside of school and more about their hopes for the future. This shift in the questions that I ask has also given me insight into their views on work and on their career goals. For example, in one questionnaire I used the prompt, "Write about your current job or life responsibilities—be specific about the type of work you do on a regular basis outside of school." One student responded to this question with: "I do UPS work which means half the time they have me packaging. One of my responsibilities is to look after my brother and sister." I believe that this type of question puts the whole person at the center, acknowledging that school is just one of the many things going on in their lives.

Another question that I asked was, "How is school helping you or not helping you in your personal responsibilities and in reaching your career goals? Be specific (it is OK if you feel that it is not helping you but write why not)." The following response gave me insight into the life of a student who rarely attended class:

School is helping me become more responsible but I'm already responsible enough without it. At this point high school is not something that is helping me currently. I honestly feel like it is slowing me down because the work is easy. I just have a problem going to school. The only reason I'm still going is because I want to graduate and make it to college where then I'm taking the next step into my life.

This response is from a student who transferred to our school and began attending during the pandemic, and was written after we had returned to physically being in person. His attendance was low even with the opportunity to work from home. He continued to attend class sporadically, but the honesty in this response helped us to begin a relational connection aimed at his specific concerns. I now have some insight into this student's thinking, so I am able to encourage him to view short-term actions toward graduation within the context of his long-term goals. Each time I see him we deepen the conversation and I learn more about him, including that he is already earning income through working with an older sibling.

Learning about my students' interests, lived experiences, and concerns with school allows me to differentiate content toward areas of greater interest, and also helps them learn to network. As a transfer school, our students have access to a Learning to Work program where they may participate in paid internships and other job readiness and college and career exploration activities. This student told me that he never saw value in the program since he is already earning an income. However, I have been able to discuss with him how he may start to build a network of mentors through an internship. The relationship insight also allows me to help students connect with other teachers who have deeper knowledge about topics such as financial investing, real estate, and building entrepreneurial skills. I have come to believe that it is critical to help students see that time in school can be useful in working toward achieving their personal and professional goals.

Relationship Building

Building relationships cannot be forced. Teaching in a transfer school, I have come to learn that relationship building does not mean bartering for better behavior or compliance. Relationships are real and take time. The building blocks include giving students time to test their voice by advocating for themselves, even if it begins as confrontational. Taking cues from question-naires and one-on-one discussions, I have found that being genuinely who I am as I learn more about students' goals and perspectives helps to build mutual respect. As I seek ways to get to know my students, they are also testing me. I continue to learn how to react and respond to a myriad of behaviors. This is often referred to as a kind of trade-off analysis of "choosing your battles," but I am also learning that "battle" is not a helpful concept; rather, I prefer to think of it as "choosing my relational methods."

Relationship-building is a two-way street. Just as in all aspects of life, relationships are built both ways, and not all students are receptive. Does this mean I am not practicing culturally responsive pedagogy effectively? Not necessarily. Not all students will like us. Not all people we meet in life will like us. However, as part of my growth and concern for my students, I may need to modify my approach.

What I learn through my relationships with students helps when I hear statements like "but I am still getting my work done" when clearly there is not a focus on the content. For me, these statements can be tied to how students perceive value in a learning activity and are a reminder that I may need to reflect on my instruction. Interestingly, these also open up opportunities to build deeper relationships by stopping to discuss the concept of work within the larger goals of doing science. It also allows opportunities to discuss more about what interests the students and about their future goals. My curiosity about my students is genuine and I know this factors into success for everyone. Building relationships is also a means for me to differentiate for each student as I learn what may be hampering access to the content.

Reciprocal Storytelling

We have all come into teaching in different ways and students want to know us and why we choose to teach. Just as I am curious about my students, they are curious about me, and ask, "Miss, why are you here?" This reciprocation of our stories is important. I tell them, "We are all on a journey, and for me, high school felt stressful. I was not always the best student, so I am paying a debt to teachers who were mentors and friends who helped me to focus academically. Without them, I am not sure where I would have ended up."

I share my own life stories with students, such as how I struggled in subjects like math, statistics, and economics. Community college made up for some of what I did not get in high school. I was inspired by the science I watched on television, like *The Undersea World of Jacques Cousteau*, *The Wild Kingdom*, and *Star Trek*. Wanting to explore the "undersea world" got me to a state university studying for a biology degree. Biology morphed into engineering with my realization that to go where Cousteau had gone I would need a Ph.D., but I was a young woman lacking academic confidence. My students enjoy these conversations and they are great since they may have heard of *Star Trek* but never of Jacques Cousteau, thus allowing for additional ties into science history.

In addition, my experiences provide an opportunity to discuss how community college can be a way forward toward a four-year college and a potential path to consider. I also share that I am a lifelong learner and that education ultimately opened doors for me that I could not have imagined during my years of waitressing, busing tables, washing dishes, retail sales, and factory work to pay my way through my undergraduate degree. I studied engineering and worked in that field for many years before becoming a teacher. I can leverage their curiosity about me to find out what they know about engineering, and about career goals they may have, and use this information to differentiate science content.

It is also important for students to realize that it is okay not to be sure about what they want to do after high school. I share that I had ideas about a future in some type of biology. As I progressed in college, I found that there were supports in place within the engineering degree path that were more robust than those in biology. It was like I accidentally ended up being drawn to engineering because of my financial situation. Students worry about paying for college and I can share that many colleges and universities have a cooperative engineering program that provides placement in a professional engineering practice with pay and academic credit. I benefited from the opportunity to experience a professional working environment; in fact, it was transformative for me. I met many women who had not seen entrée into engineering but who paved a way for me to ultimately do so. It is important that students hear these life stories from us and that we incorporate readings and videos that highlight both the diversity of people and experiences behind the science content we are teaching.

I acknowledge to my students that my formative experiences pale in contrast to the societal inequities they may experience, but my life experiences influence my thinking and my teaching. This awareness reminds me to incorporate time to talk with my students about their experiences and improves my empathy skills, while helping my students build confidence toward their goals. These conversations help provide background as I incorporate examples of jobs and job progressions that can lead to a science-based career. I am able to connect those who do not want to go to college to ideas for positioning themselves with options and career opportunities, such as auto mechanic trades, that can lead to an associate's degree in mechanical engineering. Cosmetology and studies to become an esthetician are areas of interest for many of my students and I work to incorporate these interests into my science teaching and discussions as well, such as integrating information about minerals in cosmetics when teaching about rocks and minerals.

I acknowledge that every school setting and classroom is different and it can seem daunting with 34 students per class to share our stories. However, relationships are reciprocal and as we exchange life stories with our students we build opportunities for them to gain insight into career planning.

Through these examples of how I take information from check-in questions, discussions, questionnaires, and other activities, and use what I learn to tie my students' stories and lived experiences to the science we are studying, I can show how I incorporate CRSE through ongoing relationship building within my curriculum. This brings me to relevance.

Relevance

As I have developed relationships with these young people, I have come to appreciate that many of my students see school in general, and science in particular, as irrelevant or esoteric and not practically applicable for them personally or professionally. Relationships build space for relevance to develop. When content is made relevant, space for blending academic discourse with more student-focused conversation increases. For example, one of my students who was working late at night at a recording studio mixing music became more interested in the electromagnetic spectrum as we talked about how light and sound differ and how artists work with scientists on sonification of scientific data. Students' interests are an important starting place for creating a classroom that encourages curiosity and opportunities for students to pursue science topics that can be relevant to them.

Science teachers can support the critical thinking and communications skills students will need to address social and environmental crises and the rapid changes in technology our society is facing. In addition, I want my students to gain skills that they can carry into whatever personal hopes, dreams, and goals they have for themselves. The Next Generation Science Standards (NGSS) science and engineering practices provide these foundational skills for life that will help students navigate in society (NRC, 2012).

Future Bio Poster

One way I provide real-world examples of people using science and engineering practices is by using a resource called: "This is what a scientist looks like" (The Harvard Gazette, 2020). This resource provides insight into the diversity of scientists working today. I have students create a poster of themselves modeled after the profiles of scientists featured in the online resource. The students are not required to pick a science career to profile, but rather to highlight who they are and ideas they have for their futures. This was a way, beyond a questionnaire, to have students share about themselves. My students responded well to this activity, specifically because I did not push them to pick a science career, but to choose career goals they are interested in pursuing. The website provides students with many examples and I asked them to pick at least three scientists to read about and reflect on prior to creating their own poster. As the year progresses, the students' posters become vehicles to initiate discussions that tie back to Earth science content. For example, I have several students interested in becoming electricians and several who are already working as estheticians. From this knowledge, I can include information or projects related to minerals in electronic devices and in cosmetic products.

Addressing Social Justice Issues

I also seek to find ways to incorporate examples of scientists, engineers, and community stakeholders engaging in high-stakes problem solving concerning real-world issues such as water rights, climate change impacts, and environmental restoration. By incorporating these examples into lessons and investigations, I address CRSE by adopting and supporting my students in developing a critical stance toward sociopolitical structures and processes (Ladson-Billings, 1995b; Paris & Alim, 2017). For instance, in a unit on the Flint Water Crisis, my students use the NGSS science and engineering practices to work toward a scientific understanding of what occurred and how it occurred. We then address the growing evidence of the social and political reasons the crisis was allowed to happen. Students conduct research to find current events that relate to this historic crisis and consider if their communities could be at risk. We extend this to other examples such as plastic waste and how students can play an active role in designing solutions. As I will be teaching this unit again in the future, I plan to develop other areas that students express an interest in, such as the locations of landfills and industrial sites and how they relate to rates of asthma by community. I am also interested in incorporating online data and data visualizations such as those available in the NYC Open Data Project (City of New York, 2022).

Incorporating Technology

I have come to appreciate a more comprehensive approach to teaching science called Science, Technology, Engineering, Art, and Math (STEAM), as this allows me to focus on all types of technology, including video games and live streaming. I am always learning from my students about the technologies that interest them. For instance, one student taught me about their interest in a platform called "Twitch." This student always seemed exhausted, and told me that they were trying to make a living by streaming on this platform. They felt that live streaming was what they were most interested in and that if they did not spend the time required to monetize the platform now, then the opportunity would be gone. This was a difficult conversation, because as a teacher I want to help my students balance their personal goals with completing their high school education. However, technology is moving at an ever-increasing pace and I look for ways for students to help me keep up with what is happening and how I might incorporate these emerging technologies into what we are studying.

Another way I use technology to find relevance in science class is an evidence-based activity where students write about their favorite form of social media, and then back up their claims with evidence from the internet about the uses of different media platforms. Students then exchange their writing and review each other's arguments with comments on post-it notes. We then have a whole class discussion on how they were able to use their expertise to write persuasively. Through this form of personal narrative, students become the experts, and this can lead to discussions about scientists using evidence, such as to explain plate tectonics or to understand COVID-19.

Cell phones are also an important part of youth culture, and I look for opportunities for students to use them as resources in their learning. Cell phones can be distractors, escapes, or powerful science learning tools. Most of my students have never used a paper map to navigate to a destination. They do not see the point in learning how to use a paper map when they have GPS navigation in their cell phone to do this for them, so I have been working to make the use of maps in lessons more relevant. For example, I have revised lessons about maps to incorporate students' voices and experiences. "Memory maps" is one strategy to help me do that.

Memory Maps

Having students develop "memory maps" directly incorporates students' voices and experiences into Earth science teaching (Hogan, n.d.). Students drive their learning through writing personal narratives and then making maps from memory of the places that they write about. Students may choose to make the map first and then write their narrative. We then compare their memory maps to the maps on their phones, and discuss how map making has changed over time. I make sure to note how technology has put us at the center of maps, and discuss with the students how that changes how we see the world. This can be an opportunity to begin to have students "map" their personal experiences, including how they travel to school. Students use their maps to think about what makes the city livable, such as water, electricity, and transportation (see Figure 7.2). They also consider the problems they

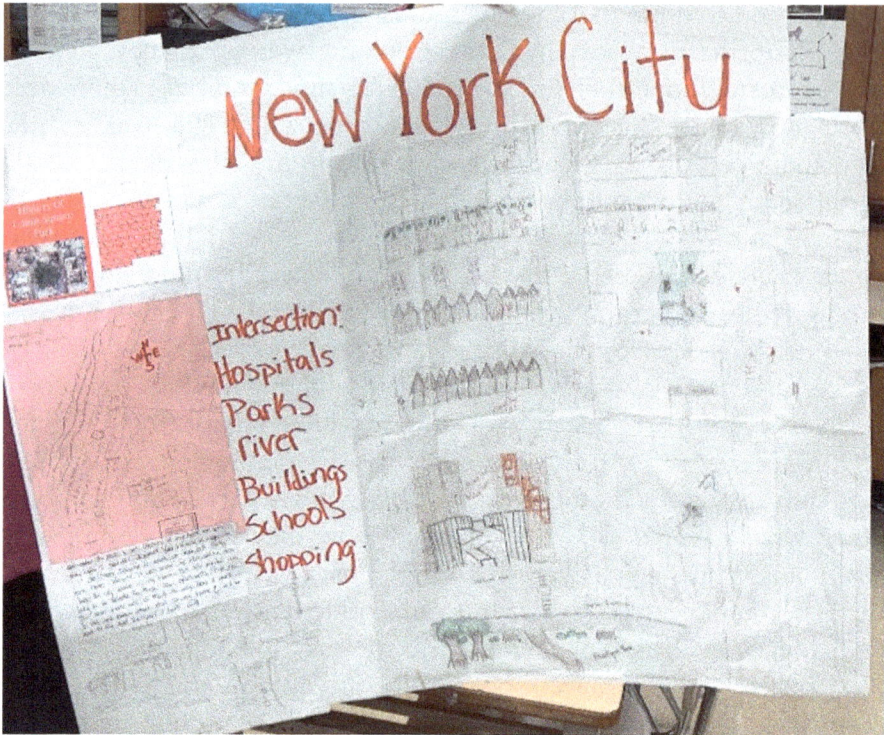

Figure 7.2 Photo of one poster that students in my science class made of New York City and intersections with infrastructure. Photograph by the author.

may encounter in their urban setting, such as higher rates of asthma or lack of affordable housing. I then can also use what I learn from my students through their memory maps to integrate their ideas and experiences into science units, particularly as they are related to students' personal goals (see sample of a student's memory map in Figure 7.3).

Maps are a type of model. In general, the use of maps and other models in science can support students' spatial orientation skills, and provide opportunities to study the city and how it was changed topographically to accommodate the modern construction and the infrastructure that the students use on a daily basis (see Figure 7.2). We also use maps as a starting point for delving into social justice issues by learning about the original inhabitants of the city, current examples of gentrification, and changes over time to communities. Connecting to social justice issues such as these is another way in which I seek to support my students in developing a critical stance.

The memory maps activity is an example of my goal to open up opportunities for students to explain something in relation to their

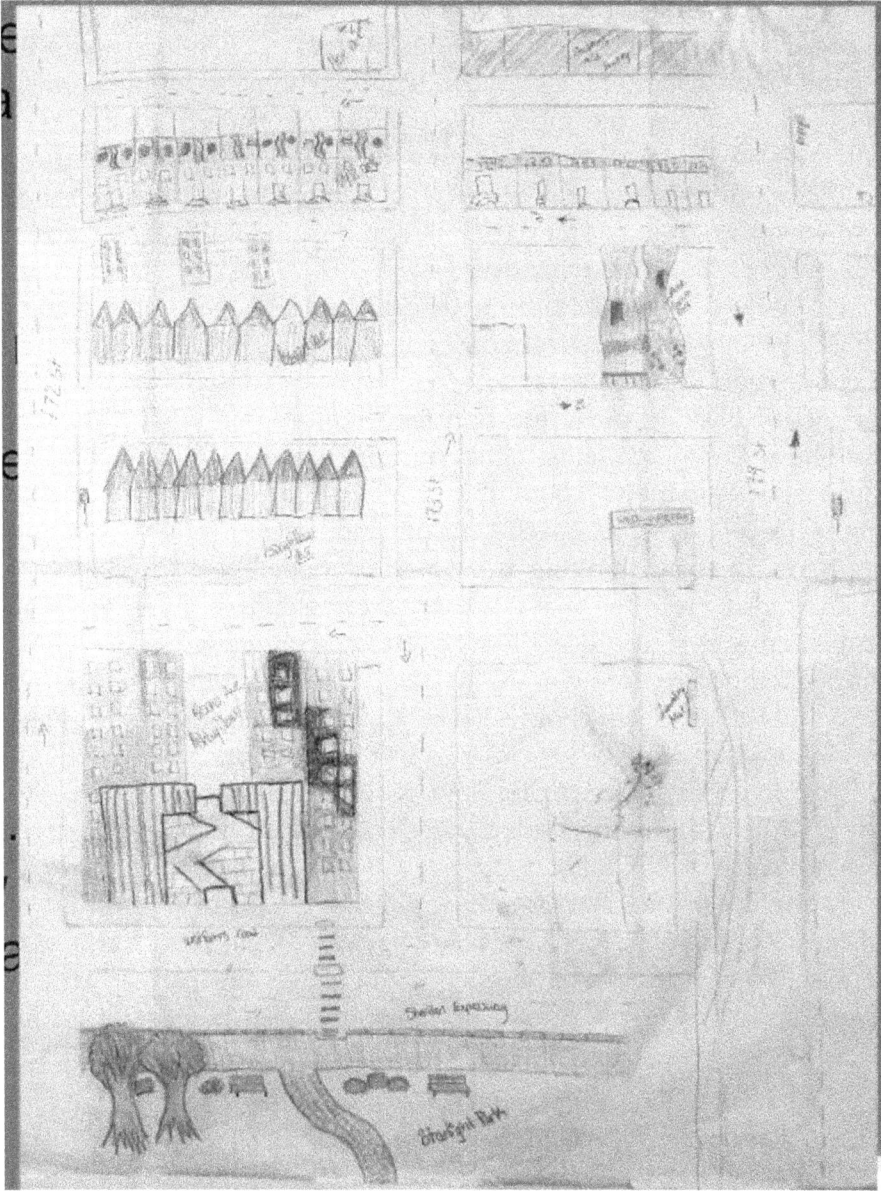

Figure 7.3 This image shows one of my student's memory maps of their neighborhood. In the image, we can see how students visualized maps of places that they navigate and represent differing student perspectives. Screenshots by the author.

personal interests, or to restate a phenomenon in their own words. Additionally, this activity highlights drawing on my students' lived experiences and communities in connection to what we are learning in science and making cultural connections, another CRSE tenet. As the school year progresses, I include check-in questions about topics students have brought up and use their questions about their neighborhoods in the curriculum. For instance, I ask my students, "When and where would you be most likely to see the moon in your neighborhood?" This can be a question about elevation, or about why you can see the moon during the day when conditions are right. Students utilize online maps to find the best spot to view the moon or stars from here in the city. I also make a point to take pictures of the moon whenever I see it during the day and ask them to do the same (see Figure 7.4). Students can then add these photos to other writings or maps of their neighborhoods.

I continue to explore ways to get to know my students and their lived experiences by asking them questions and building on what I learn. For a geographical example, many of my students are from Caribbean nations, and can speak firsthand about the effects of climate change, including rising sea levels causing beach erosion and other problems (Strauss & Kulp, 2018). In turn, thinking from a STEAM perspective allows me to broaden what I can offer to my students. For example, I can teach about the engineering design process to help students understand how it and the scientific method are iterative. In this vein, I use our collective experiences with COVID-19 to clarify how science and engineering worked together from the development of the vaccines to their manufacture and distribution. This involved collaboration, and we discussed how being part of a team pursuing a joint effort with diverse colleagues can be an empowering experience as we work together in our lab groups. I have also added art projects into my science classes to include the interests of some of my more artistically inclined students. One way we did this was to make paint from minerals, which my students then used to create paintings of the content we were studying.

Reflection

In many ways, CRSE is a process for both teachers and students. As I have explored what it means for me to be a teacher who is aware of the need to co-create CRSE experiences with my students, I have learned how important it is to reflect on my teaching practice. When students determine what is relevant and I am responsive to their ideas, then the processes of learning can become substantive. I can also build new spaces for students to express how a lesson felt for them and if a change would help them feel more engaged. The forms of reflection that I use in the classroom are exit tickets, weekly conferencing, and class discussions on how the topic or lesson is going.

Figure 7.4 Viewing the daytime moon from the NYC subway. Photograph by the author.

Reflection thus allows us to come full circle. Culturally responsive and sustaining teaching includes all students of all backgrounds. As teachers, we need to be aware that it includes us as well. We bring our own cultures and biases into all that we do and we must continuously reflect on this. I regularly ask how lessons can be modified to match my students' personal responsibilities and career goals. In addition, I ask my students to reflect upon their learning and upon my teaching. What can I change that would make content better for them? I also challenge them to make connections between learning and their personal and career goals.

I am in my fifth year of studying CRSE with the teacher professional learning group called the CRSE PLG (described earlier). Being a part of a learning group has been an opportunity for growth in my teaching. Reading and reflecting has helped me see that I have benefited from a society that privileges some above many others. It has alerted me to the fact that just as my students do not enter the classroom as "blank slates," I, too, bring life experiences and biases to my teaching. By being part of the PLG engaged in learning about CRSE, more than anything else, I sense that blinders started to fall off for me. I see more clearly the challenges many of my students face in a society that pushes a narrative different from their experiences, which helps me to see that I, too, have adopted a critical stance in addition to supporting my students in their development in this regard. While I want to be an encouragement to all my students and help them develop life skills, doing so from a place of privilege alarms me. How do I as a white woman inadvertently continue inequities? How can I acknowledge my positionality while teaching my students to experience the power of scientific thinking and amplify their voices in a way that is of maximum value to them? These are challenging questions that I continue to reflect upon as I strive to meet my students' needs as science learners, and support them in progressing toward fulfilling their dreams.

Conclusion

While I have not invented new teaching methods or a new take on CRSE, I have empirically arrived at how to be intentional in leveraging the 3Rs of relationship, relevance, and reflection to create a positive learning experience for my students. Teaching is a conscious choice. As such, I remind myself to be self-reflective and not averse to accountability. We, as teachers, need to guard ourselves against burn-out and becoming myopic out of sheer survival. We do have a high accountability to be willing to change our teaching to most benefit our students. CRSE is an opportunity for us to continue our growth. If you, like me, feel confused about the cultural part of CRSE, consider that you already have the foundation of relationship building to become the teacher that your students most need.

Reflection and Discussion Questions

1 Which of the Rs resonate most with your teaching practice in terms of enacting CRSE in your classroom?

Relationship:

2 How do you show ongoing relationship building within your daily lesson plans?

Relevance:

3 How can information from your relationship building begin to bring relevance to your science content?
4 How can we create classrooms that mirror professional workplaces?
5 How can we build choice into our curriculum where students pursue science topics of interest to them?
6 What tools do you bring to support your students in gaining skills that will help them navigate whatever they encounter in society and their working lives?

Reflection:

7 Look at one of your favorite, least favorite, or most challenging lessons. How could the lesson objective be modified to match students' personal responsibilities and career goals?
8 How do you engage your students in reflecting on your teaching?
9 What might you change that would make content better for them?
10 How might you challenge your students to make connections between learning and their personal and career goals?

Note

1 In the USA, the expected age at high-school graduation is 18.

References

Alliance for Excellent Education. (2011). *Helping students get back on track: What federal policymakers can learn from New York City's Multiple Pathways to Graduation Initiative.* https://schottfoundation.org/wp-content/uploads/HelpingStudentsNYC.pdf

Bloom, B. S., Engelhart, M. D., Furst, E. J., Hill, W. H., & Krathwohl, D. R. (1956). *Taxonomy of educational objectives: The classification of educational goals (Handbook 1: Cognitive domain)*. Longmans, Green & Co Ltd. https://ia903005.us.archive.org/15/items/bloometaltax onomyofeducationalobjectives/Bloom%20et%20al%20-Taxonomy%20of%20Educational %20Objectives.pdf

City of New York. (2022). NYC Open Data Project. Retrieved from: https://opendata. cityofnewyork.us/projects/

Danielson Group. (2022). *The framework for teaching*. https://danielsongroup.org/the-framework-for-teaching/

Good Shepherd Services. (2007). A model transfer school for replication. https:// goodshepherds.org/wp-content/uploads/2015/09/11027.pdf

Hammond, Z. (2014). *Culturally responsive teaching and the brain: Promoting authentic engagement and rigor among culturally and linguistically diverse students*. Corwin Press.

Harvard Gazette. (2020, July 23). *This is what a scientist looks like*. https://news.harvard.edu/ gazette/story/2020/07/i-am-a-scientist/

Hogan, J. (n.d.). *Memory maps*. OECD. https://www.oecd.org/education/ceri/Memory-maps.pdf

Jones, R. D. (2010). *Rigor and relevance handbook* (2nd ed.). International Center for Leadership in Education. https://www.in.gov/gwc/cte/files/ncteb-rigorrev.pdf

Ladson-Billings, G. (1995a). But that's just good teaching! The case for culturally relevant pedagogy. *Theory Into Practice, 34*(3), 159–165. 10.1080/00405849509543675

Ladson-Billings, G. (1995b). Toward a theory of culturally relevant pedagogy. *American Educational Research Journal, 32*(3), 465–491. 10.3102/00028312032003465

National Research Council (NRC). (2012). *A framework for K-12 science education: Practices, crosscutting concepts, and core ideas*. National Academies Press.

New York City Department of Education (NYC DOE). (n.d.). *NYC transfer schools*. Retrieved December 20, 2022 from https://transferhighschools.nyc/

New York State Education Department (NYSED). (2019). *Culturally responsive-sustaining education framework*. http://www.nysed.gov/crs/framework

Paris, D., & Alim, H. S. (Eds.). (2017). *Culturally sustaining pedagogies: Teaching and learning for justice in a changing world*. Teachers College Press.

Steinberg, A., & Almeida, C. A. (2010). Expanding the pathway to postsecondary success: How recuperative back-on-track schools are making a difference. *New Directions for Youth Development, 2010*(127), 87–100. 10.1002/yd.365

Strauss, B., & Kulp, S. (2018). *Sea-level rise threats in the Caribbean: Data, tools, and analysis for a more resilient future*. Inter-American Development Bank. https://sealevel.climatecentral. org/uploads/ssrf/Sea-level-rise-threats-in-the-Caribbean.pdf

Webb, N. L. (2002). Depth-of-knowledge levels for four content areas [Unpublished paper]. https://www.maine.gov/doe/sites/maine.gov.doe/files/inline-files/dok.pdf

Part

3

School-Based Teacher Collaboration Reflecting CRSE

8

Before CRSE

Trauma-Informed and Healing-Centered Strategies

Caity Tully Monahan

HIGHLIGHTS

- In this chapter, Caity considers the question: How do I use culturally responsive and sustaining and trauma-informed practices in my classroom?
- Focusing on healing-centered strategies, Caity describes how she facilitates circle practices, an activity to examine one's own identity. In doing this, she delves into the importance of exploring her own identity as well.
- Considering how to best meet the needs of students in the school, Caity shares how she and her colleagues tested an interdisciplinary studies approach to developing classes during the pandemic.
- We hope you take away from this chapter an informed understanding of healing-centered approaches and strategies to adapt for your own context.

I begin this chapter with a poem that my colleague wrote for the students in a graduating class at my school. This poem sets the stage for exploring

DOI: 10.4324/9781003397977-12

trauma-informed, healing-centered, culturally responsive and sustaining approaches that I use in my science classroom.

A Poetic Gift from beloved, distinguished and extraordinary, Mr. David Ward, to the graduating class of 2021

We See You

This moment was foretold
Centuries before your birth
Many of us
Knew as we took
The long march down familiar
Dusty roads bound together
Not only in metal bands and rope
But joined together with
Hope

We walked to the ships
To travel the paths of the hurricanes
From homeland to a God forsaken place
Unlike the pastoral plains and
Canopy of leaves that covers the
Womb of all
humankind
We arrived unaware to this new land
Uncertain of the next tomorrow
But always knowledgeable
Of you

Our legacy is not
Forged only in our endurance
Tenacity and resilience
Our testimony is not
Simply surviving the heat
The hate
The lash
Our third eye
Saw tomorrow

And it walked
In You
We knew we
Could soothe the dog bites

Mend the bruises from water canons
Overcome the indignities of being
Unseen and
Less than
Because we understood
Every lie
Must yield to
Truth
And you are
The destroyers of
Myths
Keepers of promise
The living griots of
Our stories

You have accomplished
What we only
Dreamed of

When you
Cast your gaze on the white crystals
Resting on black velvet at midnight
Know the eyes
Of your ancestors
See you
and they
Rejoice

d. alan ward
June 16, 2021

Introduction and Context

Since graduating from my master's program in 2016, I have been teaching at
the same public transfer high school in New York City (NYC). At my school,
I feel respected by my administration, my colleagues, and the students (98%
of the time). I teach a combination of Earth science and life science classes
that are ultimately assessed through a performance based assessment task
(PBAT), rather than by the statewide standardized exam. The NYC
Department of Education (DOE) describes transfer schools as "small, full-
time high schools designed to re-engage students who have dropped out or
fallen behind in credits. Transfer high schools are academically rigorous and

students earn a high school diploma" (NYC DOE, 2023). To me, transfer schools are for students who want a change in their learning environment, have been pushed out of their previous schools, or have had to leave school and are trying to re-engage.

The transfer school in which I teach serves a student population that is predominantly Black and Hispanic or Latinx, with one-third of the students having an Individualized Education Program (IEP) (NYC DOE, n.d.). Approximately 60% of the students at the school have fewer credits than most students in their age group; one of the main goals of the school is to support students in their move toward a diploma and potential college attendance.

Many of the reasons that students arrive at transfer schools are deeply personal and not always fully shared with staff or peers. Many of our students show up to their in-take meetings saying that they don't feel like they belong in the classroom, that they feel they've been unsuccessful at their previous school for whatever reason, or they "want a fresh start." Many students have experienced extreme trauma, and the stories they shared with me have informed how I teach. I am not going to share those stories here. Instead, I explore the question, *How do I use culturally responsive and sustaining, and trauma-informed practices in my science classroom?* To explore this question, I draw from instructional resources, student work, attendance and grading data, and my reflections as a teacher. I focus on how my understanding of culturally responsive and sustaining education (CRSE) has evolved over time and how this has played out in my science classroom. I also delve into the structures across my school that support CRSE and respond to the needs of our students.

Building Trust

Professor Christopher Emdin, author of *Ratchetdemic* (2021), *STEM, STEAM, Make, Dream* (2022), and my personal favorite, *For White Folks Who Teach in the Hood … and the Rest of Y'all Too* (2016), once came to speak to my graduate school class. In his talk, he said, "Kids don't learn from people they don't like." I didn't decide to go into teaching to be loved by all, so I choose to interpret this as "students don't learn from people they don't trust." My personal philosophy is that, in order to get a student to learn in my classroom, trust needs to first be built between (in no particular order): teachers and colleagues (including administrators, who also have to trust teachers), teachers and students, and students and peers.

This idea of trust building as "Learning Partnerships" was described by Hammond (2014) in *Culturally Responsive Teaching and the Brain.* Building trust with my colleagues seems like a no-brainer. For the seasoned educators reading this, you've probably been to a professional development workshop focused on building trust or friendships with your colleagues and you are rolling your eyes at reading this. However, trust does not mean that teachers need to be best friends. I think my co-teacher said it best with, "I don't have to like you. I just have to trust that you are going to do your job." Not only do institutions function better as a community with trust, but we also model trustworthy relationships for our students. This commitment can also lead to building trust with one's students.

In building trust with my students, I hope that they can further develop their identities as scholars. By "identity of a scholar" I mean that sense of identity where students can walk into a school or a classroom confidently and say affirmations such as:

- "I belong here."
- "I can learn."
- "I can earn a credit in this class."
- "Not only can I earn a credit in this class, I can challenge myself in this class and take this knowledge with me to apply it in my own life."

I believe that this latter affirmation, which connects to our CRSE tenet of high expectations, is at the heart of culturally responsive and sustaining teaching. And if there is no trust built between my students and myself, I cannot learn enough about my students to create a culturally responsive and sustaining curriculum and cannot be that culturally responsive and sustaining teacher that they deserve.

Schoolwide Structures

There are several structures implemented across my school that I feel are integral and supportive of building trust, are responsive to our students, and set a foundation for engaging in CRSE practices in the classroom. In this section, I explore two of these structures at my school: performance-based assessment tasks and goal setting.

Performance-Based Assessment Tasks

One structure that has been beneficial in building trust with my students is my school's lack of state exams. My school is part of the New York Performance Standards Consortium (hereafter, "the Consortium") of 38 public high schools that received a waiver from New York State to replace the Regents exams with another form of assessment. Instead of the standardized test, my students complete a performance-based assessment task (PBAT). A PBAT is a detailed paper written on a science experiment that the students design, carry out, and then present. It's intended to be very much like real science and it can provide that "fresh start" that our incoming students request.

In my science classes, a PBAT is a 15- to 20-page paper that is written on a project that students carry out over the course of a semester. The students present their PBATs to a panel of teachers and an outside evaluator in written form and orally. The PBAT is then scored using a complex assessment rubric developed and tweaked annually by the Consortium. Thus, the PBATs are developed by teachers and students, and assessed externally. PBATs are exceedingly difficult for some young people, because completion demands more than memorizing information and showing up with a #2 pencil. Completing a PBAT requires a lot of independence, creativity, design thinking, writing skills, and presentation confidence. Recent research suggests that PBATs support students' academic learning in a holistic and student-centered way, with the potential to enhance equity and access for students (Fine & Pryiomka, 2020).

Before I have my students begin a PBAT, I need to make sure that they feel that they belong in the classroom, that they have that identity of a scholar, and that they are willing to engage in high-level work. This is not something that is unique to the science classroom—it is felt by teachers and staff who work with our young people across the board. In order to help students learn, there must be trust.

Schoolwide Goals

In the fall of 2020, after school buildings had been closed since March due to the COVID-19 pandemic, our school community revised our instructional goals to meet the needs of our students. "Visioning," as our meeting was called, was not prompted by the pandemic, although that certainly informed the tone and scope of our goals. It is an annual meeting for our school administrators, teachers, and staff to discuss our progress and collaborate on crafting goals. In this chapter, I look at one of the goals that we established at my school: *To focus on culturally responsive and trauma-informed approaches to schoolwide and classroom structures, strategies, and differentiation in order to increase student engagement, academic achievement, and post-graduation readiness.*

In my school context, healing-centered spaces and culturally responsive and sustaining approaches must go hand-in-hand. To be clear, "trauma" is not a culture. However, there is a collective experience in my school where most students, although not all, believe for some reason that they don't belong in an academic setting. As educators, we must disrupt this with an asset-based approach that sees students for who they want to become, rather than who they believe themselves to be at the moment. After all, adopting an asset-based approach (López, 2017) is critical to CRSE and represents one of our CRSE tenets.

I, as do most teachers, want to truly "see" my students so that I can know how to best teach them; to understand their specific needs; and to create lessons that excite, engage, and are relevant for the person they want to be in this world. As Hammond (2014) puts it,

> In culturally responsive teaching, rapport is connected to the idea of affirmation. Affirmation simply means that we acknowledge the personhood of our students through words and actions that say to them, "I care about you." Too often, we confuse affirmation with building up a student's self-esteem. As educators, we think it's our job to make students of color, English learners, or poor students feel good about themselves. That's a deficit view of affirmation. In reality, most parents of culturally and linguistically diverse students do a good job of helping their children develop positive self-esteem. It is when they come to school that many students of color begin to feel marginalized, unseen, and silenced. (pp. 75–76)

Most of my students have high self-worth. They know who they are. But many have also felt marginalized, unseen, or silenced in their previous school settings. When some of my students walk through the school doors, something changes from their outside-of-school persona to their inside-of-school persona. I need to figure out how to truly see them for who they are and how they want to be seen. This way, I can understand what motivates them and will support them in engaging with high-level and rigorous science work.

It is also important to have some understanding of student triggers, especially if they are likely to come up in an academic setting. As Bessel van der Kolk describes, "Trauma produces actual physiological changes, including a recalibration of the brain's alarm system, an increase in stress hormone" (2014, pp. 2–3). As a teacher, I aim to support my students in learning science. Science teaching cannot happen if my students are not ready.

Researching Trauma

In January 2022, an article was published titled "How trauma became the word of the decade: The very real psychiatric term has become so omnipresent in pop culture that some experts worry it's losing its meaning" (Pandell, 2022). In my experience, this title is true and resonates. We throw out that word all of the time, mostly talking about events like the pandemic, quarantine, or our hunt for toilet paper in April 2020. But trauma has a very specific and clinical definition. When a young person feels intensely threatened by an event they were involved in or witnessed, that event is called trauma. van der Kolk (2014) describes trauma not as an event, but as the way the body and mind respond to an event.

There are multiple categories and types of trauma. The National Child Traumatic Stress Network (NCTSN) even has categories of child trauma that include bullying, community violence, complex trauma, disasters, early childhood trauma, intimate partner violence, medical trauma, physical abuse, refugee trauma, sexual abuse, sex trafficking, terrorism and violence, and traumatic grief (NCTSN, n.d.). Hardy and Laszloffy (2006) describe trauma in teens as a process than can lead to "devaluation," which robs a person of their dignity or self-worth; "disruption of the community," or their feeling of belonging; or "the dehumanization of loss," which can be physical, emotional, or psychological. They also mention different types of trauma using other categorizations, such as acute, chronic, collective, complex, intergenerational and historical, and secondary/vicarious (Hardy & Laszloffy, 2006). Through this typology, we see that trauma can be a single event or multiple traumatic exposures that repeat over time. In addition, a traumatic experience can affect whole groups of people, and can happen within a family or across communities, and one person's exposure to trauma can affect others with whom they interact.

Every year, our school's social workers have the teachers watch a video on adverse childhood experiences (ACEs). According to the Center for Youth Wellness (2017), a large study investigated the relationship between chronic stress caused by early adversity and long-term health. The Center for Youth Wellness explains that ACEs can build up in the body beginning in childhood and can lead to toxic stress. In essence, the more ACEs a young person experiences, the higher the likelihood that they will suffer from physical conditions such as heart disease, pulmonary issues, and diabetes. ACEs can also lead to poor academic achievement, and increased rates of substance abuse, depression, and suicidality later in life. The ACEs are categorized into three types: abuse, neglect, and household dysfunction. If you are so inclined, you can conduct a quick Google search for an 18-question quiz that National Public Radio created where you can test your personal ACE score (Starecheski, 2015). ACEs are not uncommon; in fact, 67% of adults have at least one ACE and 13% of people have had four or more ACEs (Center for Youth Wellness, 2017).

Childhood trauma isn't something you just get over as you grow up, explains Dr. Nadine Burke Harris (2015) in a TED Talk. Repeated stress of abuse, neglect, and parents struggling with mental health or substance abuse have measurable effects on the developing bodies and brains of children that they can carry into adulthood. Harris describes that exposure to early adversity can affect the nucleus accumbens, the prefrontal cortex, and the amygdala. The nucleus accumbens is typically linked to rewards and motivation; changes in the nucleus accumbens are associated with addiction. The prefrontal cortex, that part of the brain teachers are responsible for nurturing, provides cognitive control.

ACEs play out in classrooms in students' responses and triggers. We have to remember that, in addition to the more universal ACEs, there is also the ongoing collective trauma of white supremacy, racism, sexism, homo/transphobia, and more recently COVID-19 and quarantine. When all of this is combined, we get what's known as toxic stress, where young people are activated easily and often, putting their brains into fight-flight-freeze mode. And we know that the brain learns from experience, so the more you live in survival mode, the more often your brain will resort to this survival mode out of habit.

Now, what ACEs scores do NOT do is tally the number of positive experiences in a person's life. It's for this reason that school communities first need to focus on making a young person feel safe, loved, and that they belong. As many of my students have made their decision to change schools and walk the three floors up to my science classroom, they are potentially already on a healing journey.

Calling for healing centered spaces that are necessary in order to address trauma, Ginwright (2018) explains, "A healing centered approach to addressing trauma requires a different question that moves beyond 'what happened to you' to 'what's right with you' and views those exposed to trauma as agents in the creation of their own well-being rather than victims of traumatic events." In this sense, we are focusing on the entire student, not only their behaviors or responses in our classroom. One critique of "trauma informed" pedagogy is that the term reflects deficit thinking (Palma et al., 2023). Instead of focusing on someone's trauma or wondering "what happened to them," "healing centered care" is asset-based and focused on where that person is headed. In this respect, there is a potential connection with CRSE with the focus on the assets that students bring into the classroom.

Following this strand of healing centered spaces, Menakem (2021) discusses how healing is not only necessary for one student, but for a community, specifically when we are thinking about collective trauma. This is particularly challenging for me as a white, cis gender, straight teacher in a classroom of young people of color, many who identify as LGBTQIA+. I am not confident that I am doing everything right. In fact, I believe I am doing a lot of things wrong.

In a response to Ginwright's article on a healing-centered trauma-informed approach that stood out to me, Valenzuela (2021) wrote:

> I wholeheartedly believe that we must NOT push back against trauma informed care in education, rather we must demand more trauma informed pedagogy, higher quality care, qualified practitioners, and systems wide implementation. Trauma informed training cannot be facilitated by anyone who has not faced the generational, historical trauma of racism, white supremacy, hetero-patriarchy, ableism and capitalism within their own bodies, communities and practices. We must not only hyper focus on the trauma of youth of color, but also "research up" into the traumatic implications of white dominant society's historical narcissism, historical amnesia and colonialist perpetration of violence. We must be wary of trainers of trauma informed pedagogy who come from dominant identity groups, as their social position challenges their ability to be truly critical in understanding the connections between trauma, healing and social change.

I agree with Valenzuela and for this reason I want to lift up the research of individuals from non-dominant groups in this chapter. I am also sharing the work spearheaded by my administrators and social workers who are all people of Color.

I am doing my best to learn and I am working to make the safest and most rewarding learning environment for my students. Reflecting on my practice, I do this through integrating multiple strategies into my science classroom to build connections with my students, create a sense of routine and safety especially for students who are chronically absent, and to develop trust. I have incorporated community-based social-emotional learning (SEL) routines, identity concept mapping, and interdisciplinary studies. These approaches are both healing centered, as defined by Ginwright, and culturally responsive. Next, I delve into how I use each of these strategies in my science classroom.

Community-Based SEL Routines: Openers

During online teaching, I leaned into strict routines. Opener: 10 minutes. Lesson: 40 minutes. Closer: 10 minutes. Wave goodbye for a full 90 seconds.

Show them my dog. Sign off. Before the pandemic, I was never a "routine" teacher. This was mostly because I started teaching in a class with a 20% attendance rate and could never feel like a "good teacher" when class starts and I only have one student in the room, and within 20 minutes five to six more start to trickle in. I just couldn't figure it out.

During quarantine, I began to embrace routine. I started my lessons off by presenting a Google Slideshow and asking students to take a breath. Then we would have an opener: "Hello my name is ___. My pronouns are ____." This was followed by a silly icebreaker, like "Which sloth are you feeling like today?" and I would show some numbered images of sloths portraying various emotions and expressions. What I thought was going to be the biggest waste of time turned into the best use of 10 minutes ever. My quarantine classes were small. With 20 students on my roster, maybe 6–8 would show up to class each day. Towards the end of the semester, only three or four students came to class. This 10-minute opener gave students the opportunity to learn each other's names in a school to which they had just transferred and had not had an opportunity to meet anyone in person. Moreover, it got them comfortable speaking, something many of my students had rarely been asked to do at their prior schools. Almost all of my students kept their cameras off—many were sharing rooms with siblings, didn't have the resources to get their hair done, some called in from work as they needed to earn income for their families, and others just didn't feel comfortable showing their face on camera.

I also had several students who didn't feel comfortable speaking, especially at first. I gave them the option of typing their answers in the chat feature on Zoom. Ten minutes for students to be silly before starting class gave us the opportunity to get to know each other and allow them time to act like children. Students developed rapport and, because it was somewhat structured, it modeled how to interact academically in an online forum. Towards the end of the semester, when students had the option to present a PBAT, this idea of unmuting and sharing became much less scary.

Circle Practices

Circles are routines that I was first introduced to by my mentor teacher at my school. This introduction was followed by training provided through the Morningside Center for Teaching Social Responsibility. The Morningside Center is based in NYC and partners with the NYC DOE to primarily train groups of educators in SEL, racial equity, and restorative practices. They also provide classroom resources for teachers to share with their students, and help guide educators and administrators on having heavy conversations inside and outside of the classroom. I have used circles to create belonging as well as when I am having restorative conversations with a small group of students and staff.

Explaining the background and purposes of restorative circles, Marieke van Woerkom (2013) of the Morningside Center writes:

> The circle approach is grounded in a Native American philosophy and practice that values individuals as they build and maintain supportive relationships and communities. Circles are used by hundreds of tribes in North America, including the Ojibwe and Lakota. The circle process provides an alternative to the style of discussion that involves debate and challenging each other. Instead, circles create a safe and non-hierarchical place in which each person can speak without interruption. It encourages respectful listening and reflection.

Six Components to Facilitating Circle Practices

In the way that I have learned to facilitate circles, there are six components. Here I explain the six components, and briefly describe what this looks like in my classroom:

1 Everyone is seated in a circle with their body facing the inside.
2 There is a talking piece that is meaningful to the class or to the facilitator. (As an Earth science teacher, I usually used a cool rock from my rock collection.)
3 There is a centerpiece that anchors the circle. (For instance, some people share images of science content that will be learned in class, or representative images like logos, on cloth centerpieces that they place on the floor in the middle of the circle.)
4 Opening and closing ceremony.
5 Two to three rounds of questions or tasks, where members of the circle are invited but not forced to participate or share.
6 A circle facilitator. (This was usually me towards the beginning of the semester in my science classes, but I would regularly ask a student to facilitate in my advisory group.)

I have used circle practices as a nearly weekly routine in both my science classes and my advisory. (Advisory is the group of students I work with based on activities intended to support students' SEL, respect for one another, and

Figure 8.1 Students sitting in a circle in the classroom. Photograph by the author.

community building.) It is my go-to first-day lesson in science. On my centerpiece I feature images of nature, landscapes, and typically some images of interesting things that we'll learn about during our class. For example, in my astronomy class I will have close-up images from NASA of craters on Mars or of extremophiles on Earth, like tubeworms. (See Figure 8.1.)

I start off by introducing myself to the students. Next, I introduce the talking piece and explain its purpose, describe the components of the circle so that they know what to expect, and read a short poem that is relevant to the class. I will have each student pick up an image that's interesting to them. I will ask them to introduce themselves, share their names, their pronouns, how many semesters they've been at the school, what they see in the image, and why they chose it. For me, it is helpful not only to hear how every student introduces themselves and pronounces their name, but also to notice who is comfortable speaking in a group and whose body language or short answers suggest otherwise. This is important in anticipating how to structure lessons later on in the class and to see who may need help building confidence to speak or make friends.

This low-stakes task is also important because there are no "wrong answers." For many students transitioning from a more traditional school and fresh off multiple-choice testing strategies where there are clear correct and incorrect answers, this open-ended question of "describe what you see" can be difficult. Many students stumble and remark, "I don't know what this is so I can't explain it." This leads me to rephrase the prompt to, "Does it remind you of anything you've seen before?" or "How does it make you feel?" to encourage students to talk a little bit more and have the opportunity to hear their voice. The latter prompts will redirect students to a generalized syllabus and the dates of the PBAT presentations. I give students the opportunity to ask questions about the syllabus. The final prompt is about how they feel about writing a science paper and/or presenting it. The closing ceremony is usually a quote that I read about the nature of science. For example, in astronomy, I often read aloud my favorite Neil deGrasse Tyson quote: "Science needs the light of free expression to flourish" (Druyan et al., 2014).

The purpose of this first-day lesson is not to teach science content, but to create the norm that all voices deserve to be heard and that science is about questioning and thinking as opposed to memorization or knowing the correct answer. The purpose of the syllabus is to understand the expectations of the class. For students who struggle with anxiety and/or those who have experienced trauma, this is a trauma-informed practice. Being clear about when to expect big deadlines helps students anticipate and not feel blindsided by deadlines. The same is true for explaining the components of the circle. Sharing how a circle works so that young people can anticipate when they will need to speak and for how long is important for their sense of safety.

Concept Mapping

Concept maps are an alternative form of formative or summative assessment of student thinking (Ritchhart et al., 2009), which can be helpful for students who struggle with writing long science explanations. It can also serve as a tool for helping students who tend to "shut down" when asked to respond to a writing prompt.

Concept maps are one of several Thinking Routines published by Harvard's Graduate School of Education's Project Zero (2019). A concept map can be created in four steps: Generate-Sort-Connect-Elaborate (GSCE). Although many educators like to use concept maps as a diagnostic assessment to understand students' thinking on a topic or issue, I prefer to use them as a formative or summative assessment to help students improve their written work on the background sections of their PBAT. I give students colorful paper, scissors, colored pencils, and markers. I tend to use two to three hours of class time to create a concept map. If it is not a summative assessment, I encourage collaboration during the first two steps of "Generate" and "Sort."

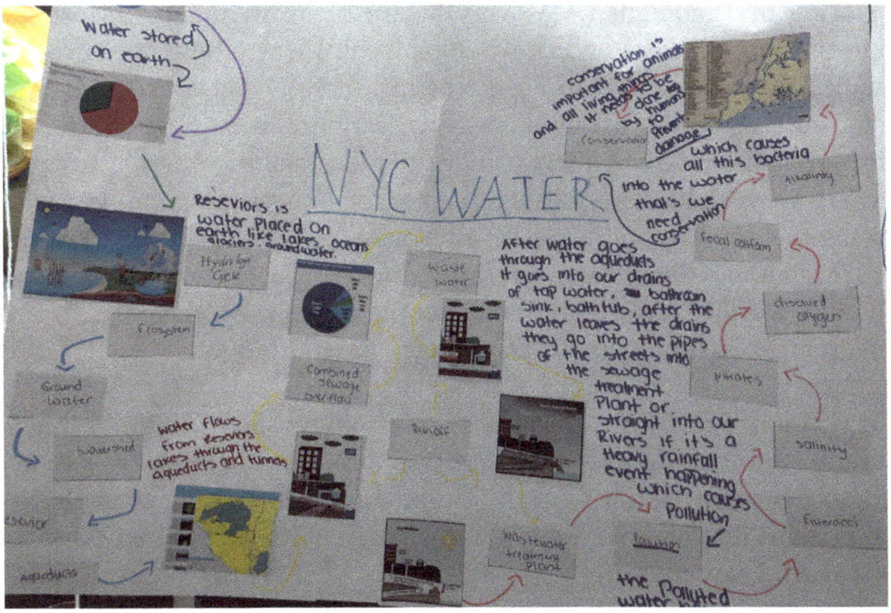

Figure 8.2 An example of a concept map on NYC water from my Earth science class. Photograph by the author.

During the third step, I ask students to walk around the room and observe others who are "Connecting." (See Figure 8.2 for an example of a concept map on NYC water developed in my science class.) The finished product is a physical representation of scientific phenomena. It tends to help students who struggle starting to write or expand on their initial ideas. Concept maps are a tool to help make students' thinking visible in an accessible way with room for creativity and individuality (Ritchhart et al., 2009). I find them to be a powerful instrument for science learning in the classroom and an alternative way to learn about my students' thinking about science concepts.

Identity Work

At the National Museum of the American Indian in New York, I participated in a two-day professional development (PD) workshop to learn about the Taíno people, heritage, and Caribbean identity. During the PD, I completed an activity to create an identity map of myself. In this icebreaker, we were asked to list the top ten words that described our identity onto different post-it notes. The workshop leaders listed examples that included ethnicity, nationality, and religion. I looked around trying to eyeball what everyone else was writing.

During that moment I was very aware of my whiteness. Most of the people around me were people of Color. Yeah I was white, but never had I thought about that as a top aspect of my identity, let alone something I wanted to brag about. My whiteness was important as it has afforded me many privileges. I have had the ability to just skate by where other people have had to work harder, question their sense of belonging, or even fear for their safety. I wrote it on the post-it anyway along with "Runner," "Athlete," and "Dog mom." Yes, those were important to me. I struggled to come up with other words. I looked around at people writing "Mother," "Sister," and then I got it. Yesssss, "Sister," "Daughter," "Granddaughter." I get this now. This is easy.

It was not. I really struggled to identify myself in ways that came very naturally to other people and this was precisely because I was a cis-gendered, straight, white woman. The identities that came strongly to me were hobbies that I enjoyed doing rather than ways that the world saw me or the ways I saw myself.

The workshop leaders encouraged us to share some of what we wrote down with the people at our table. People had written things such as "Dominican," "Gay," "Black," "Teacher," "Christian," "Auntie." We were then asked to form groups with the post-it notes of our identities out on the table. I put "Athlete" and "Runner" together, and then I placed "Daughter," "Granddaughter," and "Sister" together. I put "Reader" and "Teacher" close together as well. "Dog mom" stood alone. We were then asked to take three identities away. Easy. I took away "Teacher," "Sister," and "Athlete." Next, we were asked to remove one identity at a time until only one remained. Everyone shared their last remaining word. Most people were left holding "Wife," "Mother," "Sister," or a nationality or race marker. I was left with "Runner." Perhaps I needed to do some more digging.

This identity exercise was helpful for getting to know the fellow educators at my table. Moreover, it was helpful in sharing and considering how we want to be seen. It allowed us to dig deep. I listened as the person across from me described the various dishes her Dominican grandmother cooked and she began to get emotional. It also allowed me to hide. It allowed us to share just as much as we wanted. For me, on that day, I could share my hobbies until I was ready to engage in deeper reflection.

I have used this activity in class and have had students create concept maps of their own identities. It's a helpful activity in that it has students share as much or as little as they feel comfortable with their teacher and peers. It's also a great way to see who struggles with the task of creating identity concept maps and for whom it comes much more naturally. By having students talk about themselves first, it allowed me to see in which steps in the Generate-Sort-Connect-Elaborate task students had initial conceptions. This also provided a way for me to get to know my students better. I was able to plan lessons that were relevant to the interests, cultures, and nationalities that my students chose to share. I asked students if they felt comfortable hanging up their maps on the walls in an area in the classroom. Posting the maps on the wall functions as a reminder that they are seen. It also ended up being tremendously useful for me, because I always looked at them when planning lessons so that I could thoughtfully and intentionally integrate my students' assets into instruction.

A Move to Interdisciplinary Studies: Testing a New Method to Better Serve Our Students

During the beginning of the pandemic, when we were told that we would be teaching remotely until the end of the school year after lockdown, we transitioned to an interdisciplinary studies (IDS) teaching method. IDS was a new approach for our school and we were eager to find out if and how it could work for our students.

We moved quickly to develop three IDS classes that spanned science, math, art, and global studies content. All three classes were connected through the topic of climate change. On Monday, we would have Climate Science and Art, Wednesday was Climate Science Statistics, and on Friday we had Climate Resilience Global Studies. Together, the IDS classes offered one credit of physical science, one credit of art, two credits of advanced math, and two credits of global studies (for a total of six credits). We were hopeful that this new approach of teaching could help strengthen connections between and across subjects, and be compelling for our students.

On Mondays, we would focus on Climate Science (Earth science) and Art. This is a class for which I was the lead teacher. The math and social studies teachers led the other two IDS classes, with support from the special services teacher. While lead teachers taught, the other teachers helped facilitate by creating and leading breakout rooms and supporting students in completing their work. Our Climate Science and Art IDS class was scheduled to meet three times a week for two hours at a time. Students had the option of joining morning or afternoon tutorial sessions for one-on-one or small group support. We had 23 students enrolled in our full cohort, five of whom were considered long-term absent (LTA), meaning that they had not attended any of the classes for any extent of time. At the end of the year, we explored how our students did in terms of earning credit in IDS classes and about their experiences.

What Did We Learn from Testing This New Approach?

After piloting these three IDS classes, we wanted to see if this approach was conducive to meeting the needs of our students. For instance, we hoped that the IDS approach would support students in earning credits. To help contextualize the findings—prior to starting the IDS classes, one student (out of the 23 registered) had earned all of their credits the previous semester. We found that, for the IDS classes, of the 18 students who attended online class at least once, all earned full science credit. Additionally, 15 students earned art credit, 14 students earned advanced math credit, and 15 students earned global studies credits. Given that all the participating students in the IDS earned full science credit, and most earned credits in other IDS subjects as well, this was exceptional. A total of 13 students earned all of their IDS credits that semester. Additionally, four students who had previously been unsuccessful at this school in terms of earning credits were able to earn all six credits. (See Table 8.1 with a tally of credits earned through IDS.) Thus, we considered the IDS in terms of credit earning a huge success!

On the last day of class, we collected feedback from our students about their experiences with IDS. Across the responses, we noticed that students

Table 8.1 Table of credits earned in IDS classes across the full cohort of 18 students who each attended class at least once. Students' names have been replaced with numbers for anonymity. Table developed by the author.

Students In Full Cohort (no LTAs)	Science Credit	Art Credit	Math Credit	Global Credit
Student 1	1	1	1	1
Student 2	1	1	1	1
Student 3	1	1	1	1
Student 4	1	1	1	1
Student 5	1	1	1	1
Student 6	1	1	1	1
Student 7	1	1	1	1
Student 8	1	1	1	1
Student 9	1	0	0	0
Student 10	1	0	1	0
Student 11	1	1	0	1
Student 12	1	1	1	1
Student 13	1	1	1	1
Student 14	1	1	0	1
Student 15	1	1	1	1
Student 16	1	0	0	0
Student 17	1	1	1	1
Student 18	1	1	1	1
Total	18	15	14	15

highlighted positively that the class subjects were connected. For instance, one student explained:

> I like that my classes were all connected in some way, because you'll always be caught up somehow. I liked it because even though they're related they have some differences, which makes it feel like it's learning something new but it's informational. Yes, it was clear which class I was being graded for. It had everyone in one stable class.

Additionally, students noted that it was helpful to have multiple teachers in the class to offer support. A few students remarked on the supportive environment in the IDS. For instance, they shared, "It's so easy to have this as one class because if you need help with any assignment, they are all there to help," and "The class is very supportive and if you're nervous, the teachers and classmates will encourage you to do it. I've experienced that situation before and it feels very good." Another student connected this support to their well-being, sharing, "It is very helpful in many ways. They make sure that your mental health is on track. They don't want you to feel overwhelmed or stressed out because of the work and the teachers will help [with] no hesitation."

Students also shared that having courses connected through an IDS approach offered an ease to completing work, locating assignments, and understanding the content. For instance, one student commented, "I felt like it was better because it was just so much easier to understand the work instead of having to focus on four different subjects but just different parts of the one subject. I was able to get my work done at a higher level because I was able to understand the work a lot better." Finally, students noted the usefulness of having the opportunity to earn multiple credits in a course, which helped them move closer to graduating (Table 8.1).

Conclusion

In this chapter, I explored various strategies and approaches to culturally responsive and healing-centered work in my science classroom. I also described structures that were set up at my school (e.g., performance-based assessment tasks and schoolwide goals) that were important in supporting this kind of work in building trust, exploring identity, engaging in restorative circles and SEL routines, and trying out new interdisciplinary teaching methods in hopes of better supporting my students. In this work, positioning students at the center is critical and involves establishing trust first, learning about their assets, creating safe spaces, and supporting them in achieving academic success. Also critical is the need for teachers to engage in self reflection, identity work, and ongoing learning at the same time.

> **Reflection and Discussion Questions**
> 1 What, if any, healing-centered or trauma-informed practices do you use with your students? In what ways do you think they connect with CRSE?
> 2 How do you and your students engage in identity work in your classroom? If you haven't tried this yet, what could you envision doing and what might that look like?
> 3 If you were interested in creating an IDS class from a CRSE perspective, what focus topic would you choose? How might you engage your colleagues in co-creating an IDS class together?

References

Center for Youth Wellness. (2017). The science. https://centerforyouthwellness.org/the-science/

Druyan, A. (Writer), Soter, S. (Writer), & Pope, B. (Director). (2014, April 7). Hiding in the light (Episode 5) [TV mini-series episode]. In S. MacFarlane, A. Druyan, B. Braga, & M. Cannold (Executive Producers), *Cosmos: A spacetime odyssey*. National Geographic Channel. Cosmos Studios; Fuzzy Door Productions; Santa Fe Studios.

Emdin, C. (2016). *For White Folks Who Teach in the Hood … and the Rest of Y'all Too: Reality Pedagogy and Urban Education*. Beacon Press.

Emdin, C. (2021). *Ratchetdemic: Reimagining Academic Success*. Beacon Press.

Emdin, C. (2022). *STEM, STEAM, Make, Dream: Reimagining the Culture of Science, Technology, Engineering, and Mathematics*. International Center for Leadership in Education, Incorporated.

Fine, M., & Pryiomka, K. (2020). *Assessing college readiness through authentic student work: How the City University of New York and the New York Performance Standards Consortium are collaborating toward equity*. Learning Policy Institute. https://learningpolicyinstitute.org/media/423/download?inline&file=RCA_CUNY_Assessing_College_Readiness_REPORT.pdf

Ginwright, S. (2018, May 31). *The future of healing: Shifting from trauma informed care to healing centered engagement*. Medium. https://ginwright.medium.com/the-future-of-healing-shifting-from-trauma-informed-care-to-healing-centered-engagement-634f557ce69c

Hammond, Z. (2014). *Culturally Responsive Teaching and the Brain: Promoting Authentic Engagement and Rigor among Culturally and Linguistically Diverse Students*. Corwin Press.

Hardy, K. V., & Laszloffy, T. A. (2006). *Teens who hurt: Clinical interventions to break the cycle of adolescent violence*. Guilford Press.

Harris, N. B. (2015). *How childhood affects health across a lifetime*. TEDEd. https://ed.ted.com/lessons/eczPoVp6

López, F. A. (2017). Altering the trajectory of the self-fulfilling prophecy: Asset-based pedagogy and classroom dynamics. *Journal of Teacher Education, 68*(2), 193–212. 10.1177/0022487116685751

Menakem, R. (2021). *My grandmother's hands: Racialized trauma and the pathway to mending our hearts and bodies*. Penguin Book Limited.

The National Child Traumatic Stress Network (NCTSN). (n.d.) Trauma types. https://www.nctsn.org/what-is-child-trauma/trauma-types

New York City Department of Education (NYC DOE). (2023). Transfer high schools. https://www.schools.nyc.gov/enrollment/other-ways-to-graduate/transfer-high-schools

New York City Department of Education (NYC DOE). (n.d.). *School quality snapshot 2021-22*. https://tools.nycenet.edu/snapshot/2022/

Palma, C., Abdou, A. S., Danforth, S., & Griffiths, A. J. (2023). Are deficit perspectives thriving in trauma-informed schools? A historical and anti-racist reflection. *Equity & Excellence in Education*. 10.1080/10665684.2023.2192983

Pandell, L. (2022, January 25). *How trauma became the word of the decade: The very real psychiatric term has become so omnipresent in pop culture that some experts worry it's losing its meaning*. Vox. https://www.vox.com/the-highlight/22876522/trauma-covid-word-origin-mental-health

Project Zero. (2019). *Generate-Sort-Connect-Elaborate*. Harvard Graduate School of Education. https://pz.harvard.edu/sites/default/files/Generate-Sort-Connect-Elaborate_0.pdf

Ritchhart, R., Turner, T., & Hadar, L. (2009). Uncovering students' thinking about thinking using concept maps. *Metacognition and Learning, 4*(2), 145–159. 10.1007/s11409-009-9040-x

Starecheski, L. (2015, March 2). *Take the ACE quiz – and learn what it does and doesn't mean*. NPR. https://www.npr.org/sections/health-shots/2015/03/02/387007941/take-the-ace-quiz-and-learn-what-it-does-and-doesnt-mean

Valenzuela, C. (2021, May 16). *Social justice education needs trauma informed care, now more than ever: A response to "The future of healing: Shifting from trauma informed care to healing centered engagement" by Dr. Shawn Ginwright*. Medium. https://candicerose.medium.com/social-justice-education-needs-trauma-informed-care-now-more-than-ever-a-response-to-healing-16706cfcb96e

van der Kolk, B. (2014). *The body keeps the score: Brain, mind, and body in the healing of trauma*. Penguin Books.

van Woerkom, M. (2013, March 3). *An introduction to circles*. Morningside Center for Teaching Social Responsibility. https://www.morningsidecenter.org/teachable-moment/lessons/introduction-circles

9

Crisis Precipitates Change

An Approach to Culturally Responsive and Sustaining Teaching Using a Modified Flipped Classroom

Arthur W. Funk

HIGHLIGHTS

- Determined to learn from pedagogical approaches implemented during the pandemic, Arthur explores his teaching practice and teacher team discussions at his school using meeting notes, teaching artifacts, journaling, and reflection. He examines the questions: *What do teachers learn about testing a flipped classroom approach during a pandemic? How do I modify the flipped classroom approach for my students in my classroom?*
- This chapter focuses on the culturally responsive and sustaining tenets of holding high expectations for students' learning and valuing students' assets.
- A teacher leader at his school, Arthur shines a light on teacher collaborative work to center students in instruction in effort to "not go back to the way things were before" the COVID-19 pandemic.
- We hope that you find in this chapter inspiration to reflect on your own practice during the pandemic and to consider what changes you have kept and why.

DOI: 10.4324/9781003397977-13

Introduction: Describing the Pandemic Context

The pandemic has had a profound impact on my growth as an educator, motivating me to improve and teach in ways I never thought possible. Not only has it necessitated rapid fluency in new platforms and technologies, but it has compelled me to reflect and reconsider my roles and responsibilities as a teacher. The pandemic continues to be a driving force behind philosophical change that is still manifesting from its disruptive inception. The pandemic has sparked a profound metamorphosis in my local educational landscape—a process in which I, as an educator, am both a participant and an observer. The changes initiated by this crisis continue to unfold, reshaping my understanding and practice of education in ways that promise to have lasting impact.

I have noticed that many of the changes brought about by the pandemic are not only useful, but essential to improving my practice. This has led me to pursue my research questions: *What do teachers learn about testing a flipped classroom approach during a pandemic? How do I modify the flipped classroom approach for my students in my classroom?* I have used multiple sources of data including teacher team meeting notes, teacher artifacts, and my biweekly journal reflections. The bulk of the data were collected during the 2020–2021 school year. Drawing on research, I present an evolution of my thinking as an extension of the voices of my school community, including the perspectives of both educators and students.

As a teacher, my ultimate goal is to equip my students with the skills they need to become independent learners. I consider independent learners to be those who take responsibility for their own learning, make choices and set goals, monitor their own progress, self-assess, and make changes based on that self-assessment. This notion is not centered around teachers being supplanted by technology. This commitment entails a deliberate strategy to limit my role as a teacher over time by guiding students towards progressively higher levels of independence.

In order to achieve this goal, I have focused on implementing teaching strategies designed with culturally responsive and sustaining education (CRSE) in mind. Through planning and reflection, I have also worked with my colleagues to give my students more autonomy and choice in their class work, allowing them to develop the independent learning skills they need to be successful in the future. In this chapter, I share reflections from our teaching team, the specific changes I have made, how I think they have affected my students' learning, and my own growth as an educator.

School Context

As an Earth science teacher in an urban, arts high school, I teach students who are predominantly Hispanic and African American, with minority enrollment making up 98% of the student body (NYC DOE, n.d.). The school has an open enrollment policy and serves the community in which it is located, but as a specialized school it also accepts outside applicants. Nearly all of my students are economically disadvantaged. Most of my students come from families that understand education has the power to change this. The school has enjoyed a history of success and garnered a positive reputation over the years. Accolades and expanding enrollment both indicate we have been doing something right. The attrition rate is low, and graduation is high. Students enjoy being in our building; many come early and stay late. I respect and enjoy working with my colleagues, and I especially love my students.

The Crisis

If you asked any teacher in my school before the pandemic, they would confirm that ours, like many other schools, had structures and practices that went against what we know about the science of learning. Despite this, many schools still endorsed these practices (Hammond, 2021). The over-reliance on standardized tests as a primary measure of student achievement was still used. Uniform curriculum was still being taught to all students, without sufficient consideration of their unique learning preferences and needs (Yang et al., 2019).

The closure of schools and the sudden transition to online learning amplified disparities and inequalities, especially regarding marginalized groups. One of the prominent challenges made more visible during the pandemic was the digital divide, which has widened the opportunity gap for many; additionally, we saw a decline in students' mental health (Milner, 2021). Underserved students already had an uphill battle, and even more cards were now being dealt against them. But to avoid painting with too large a brush, I will focus on my own teaching and my school community where, at the outset of 2020, the integration of technology and digital resources was limited and social-emotional learning was absent from the vernacular. The pandemic changed this, and inspired us to try and change a lot of other things, too.

One thing that didn't change with the pandemic was the persistent sense of fatigue. There have never been enough hours in the day, and remote instruction only added another hole to the leaky bucket of teachers' time. More stress, more work, and more tools to learn left us living like double-lit wicks. We found ourselves caught between the encroaching flames of weariness and the mounting demands of our roles. Maintaining a delicate equilibrium became critical. We needed to balance the relentless pace of change with the preservation of our own well-being.

I witnessed firsthand the shortcomings of not adequately equipping my students for remote learning. Eventually I figured out how to plan for asynchronous lessons, but I needed more time to develop content for remote learning and in-person time with my students to teach them the skills they needed to learn independently. Neither of which were possible. As a result, most of my students encountered difficulties with the competencies and routines required for self-directed work. They struggled with managing their time, keeping themselves motivated, and maintaining self-discipline—each of which are valuable for excelling in the remote learning environment. I realized that I had not taught my students how to survive without the physical presence and immediate support of a teacher. Instead, I had trained them to acquiesce and rely on my guidance. I rewarded them for advocating for assistance, but I did not teach them how to take initiative. Unintentionally, I was promoting dependent learners who did not get adequate support to facilitate their cognitive growth and who were not able to activate their own neuroplasticity (Hammond, 2014). Essentially, I had shown students how to go through the motions of learning in the presence of an adult, but I had not reinforced behaviors that would enable them to pick up where we left off in class and push forward independently.

The sudden shift to remote teaching caught educators totally off guard, making it nearly impossible to properly acclimate, adapt, and provide good instruction. The lack of preparation, the absence of social interaction, and the sense of isolation that comes with remote teaching coupled with disparate access to technology, internet, or even quiet workspaces were a few of the obstacles to student success. The abrupt amputation of in-person instruction truncated the school year and the only tourniquets to prevent total exsanguination were Zoom rooms strapped to a sloppy splint of asynchronous assignments. And yet, there were some students who did succeed. In my experience, the students who were already academically successful usually continued to do well; those who were struggling did worse, and those with special needs and individualized education programs (IEPs) suffered most due to the lack of specialized support and accommodations. Nevertheless, most of my students were either engaged or completely absent. I did not see a middle ground. A handful were ready to completely freelance high school at home, while others were ready to drop out of school entirely with no way for me to reach them.

My Positionality

As a middle-aged, cisgender, able-bodied, white science teacher, I am aware of the unearned privileges that accompany my racial identity. It is crucial for me to acknowledge the advantages that have contributed to my personal achievements, how the dynamics of power and privilege come into play, and how they might affect my interactions and relationships in the classroom. Students require role models in science from individuals who reflect their own identities (Chowning et al., 2022).

Although I do not share the same racial identity as the students I teach, I understand that, as a science educator, I need to consciously provide a safe environment that acknowledges diversity as essential to advancing science as a whole (Hofstra et al., 2020). The focus on science as an accumulation of facts creates a charade that this inclusion is somehow irrelevant, and ignores that science is a collaborative endeavor that benefits from inclusion, equity, and criticality (Godsil, 2016).

Through self-awareness, I strive to approach my teaching with humility, empathy, and a commitment to fostering a culturally responsive and inclusive learning environment. I recognize the need to continually educate myself and actively dismantle the barriers that perpetuate inequality within the education system. It is my responsibility to engage in critical self-reflection; listen to the voices and experiences of my students and their communities; and work towards promoting justice, equity, and the holistic development of every learner.

I also believe that CRSE must be designed to incorporate reflective practices that enhance self-awareness. Young learners must recognize their own biases and assumptions in order to challenge them. Recognizing one's own perspective is important in understanding the perspectives of others and building stronger relationships across cultural differences.

The Way I See CRSE in Science

In my view, culturally responsive and sustaining education (CRSE) is a philosophy that places significant value on integrating a variety of cultural perspectives into teaching and learning. This connects to valuing students' assets, an aspect of one of our CRSE tenets (López, 2017). In this chapter, I focus on the tenets of valuing what students bring to the classroom as assets, as well as holding high expectations for all students' academic learning (Ladson-Billings, 1995; Gay, 2010).

Creating a caring and supportive atmosphere that honors diverse identities is crucial for promoting the practices that drive progress in science. Failure to include all individuals is harmful to the exploration of new scientific insights and to the individuals themselves. Explicitly embracing inclusivity highlights the worth of diversity, recognizing it as an asset that enhances the potential for scientific advancement (Bali, 2015). As a science teacher committed to nurturing scientific progress, it is important to prioritize broadening participation in science. This commitment starts in the classroom.

A safe and inclusive learning environment, where all students feel valued and supported, is a prerequisite for this inclusion (NYSED, 2019). By doing so, it is my intention and belief that our young scientists will share more about their own experiences, practice and develop a sense of belonging within the field of science, and be empowered to take ownership of their learning. Before the pandemic, this was much less obvious to me. I considered CRSE an addition to what I was already doing rather than an essential philosophy, and I spent the majority of planning focused on building content knowledge. Sure, I included thoughtful, hands-on learning activities, but due to time constraints, I opted for standardized test preparation. I inadvertently prioritized students being on task, leading to a pedagogy of compliance (Delpit, 2019; Haberman, 2010; Hammond, 2021). I wanted to do more for students' confidence, independence, and self-assuredness in their learning.

As my colleagues and I transitioned from remote learning back to the classroom, we practiced caution in the implementation of strategies meant to address the diverse needs of students that might instead exacerbate existing inequities. As Hammond (2021) pointed out, "Too often, when we deem students behind academically, we increase compliance measures and actually decelerate learning. We over-scaffold rather than coach students to engage in productive struggle to process the content." Hammond's assertion critically examines this pitfall in education, revealing an important insight about the potential negative effects of imposing too much structure in the learning process. Ideally, scaffolding is gradually removed as the student becomes more proficient; however, Hammond highlights a tendency to overuse scaffolding. This leads to over-reliance on the teacher and a decrease in a student's autonomy and self-confidence. It limits the opportunities for students to engage in the kind of challenges that are manageable and yet pushes them to extend themselves. When we remove opportunities for struggle, we contradict our purpose and impede student independence.

Teacher Collaboration and Planning at School during the Pandemic

The pandemic exposed us to the voids and challenges of the system in which we operate, many of which we still face. Nevertheless, I learned a lot along with other teachers at my school when our practice necessarily shifted as the crisis precipitated changes in our instruction. Working to adjust to our students' needs, we sustained a veneer of optimism over uncertainty. We sought silver linings to reassure that our efforts were not in vain. This was made possible by school leadership, who understood our predicament and gave us needed time and space.

We did two important things across the entire staff. First, we adjusted our schedules. We scheduled daily office hours to support struggling students, and the majority of Mondays and Fridays to meet together without students. We dedicated time to collective teacher planning and synchronous instruction. Alternating between grade and department teams, every teacher meeting included time to check in, show gratitude, and bond. Second, we dramatically reduced class sizes and scheduled only one synchronous class per subject per week. Lessons were then intensively planned, and repeated multiple times throughout the entire week. This new format relied almost entirely on asynchronous learning and emphasized culminating performance tasks over traditional assessment. Live class time then prioritized checking in and getting students talking. We practiced running our teacher team meetings as we would run our online classes, and realized that if we only got to see most of our students for one hour each week, the time together really had to count.

This plan felt right, but was like nothing we had done before, and it was challenging for teachers to adjust. We continued meeting to offer each other support and share our experiences. Although we had to dig deep at times, every teacher team meeting included conversations on what we were doing that was actually working. During one of these check-ins, Bettina Love's (2020) article, "Teachers, We Cannot Go Back to the Way Things Were," became a recurring reference in our discourse. Love's article motivated our cause at a time when we needed it most. It highlighted the need for us to embrace an approach to teaching and learning that goes beyond the traditional classroom model. We foresaw the dangers of returning to business as usual after the pandemic, and focused our collaborative efforts on how we might embrace a more transformative approach to teaching.

The weeks that followed our exposure to the Love article, we kept coming back to her argument, and similar conversations bounce off the walls even now. During these meetings, I journaled about my experience as we collaborated and reflected on our work together. My notes focused on how we were addressing students' needs and building students' capacity to become independent learners. To me, the most poignant conversations happened while teachers reflected on their practice and responded to the following questions:

- What are we doing better as a result of this crisis?
- What (constructive) permanent changes to instruction could/should happen as a result of this crisis?

Within the context of one team meeting, 34 teachers recorded their initial responses to these two questions on a community Jamboard (see Figure 9.1).

During the following weeks, teachers met in department and grade teams, and several times as an entire staff to discuss our Jamboard responses to these two questions. This gave us the time and shared space to digest the complexities of our situation so we could plan how to keep improving. As it turned out, the answers to both of these questions inevitably evolved until they were no longer separate. It seems obvious, but if we were doing something better, there was no reason to stop. During our final meeting of

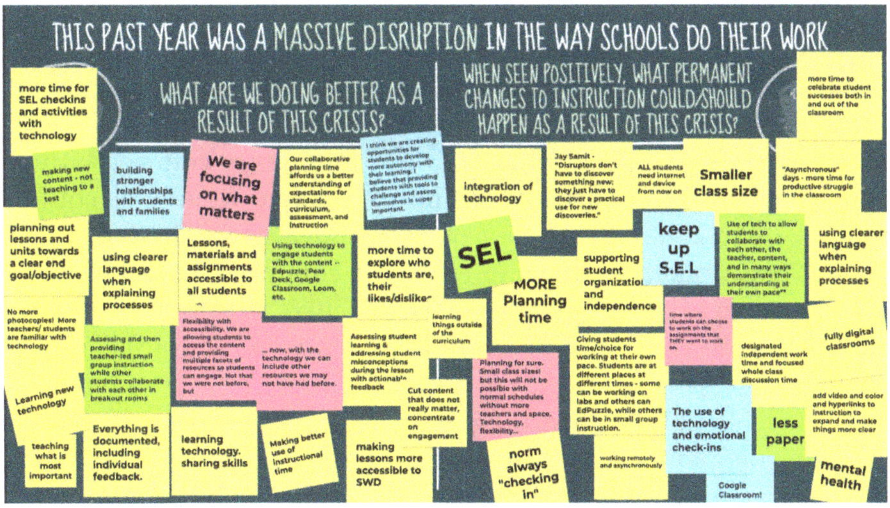

Figure 9.1 A screenshot of a Jamboard of teacher responses to two questions developed by the author and the school's Assistant Principal. Screenshot taken by the author.

the year, we discussed these questions one last time. The final responses were informed by the positives, the challenges, and what we learned. Next, I share insights derived from a compiled summary of these teacher team discussions.

Positive Insights from Teacher Team Discussions

Teachers shared many positive comments, but what stood out to me most was the pride in how we embraced social-emotional learning. My colleagues expressed that we worked hard to acknowledge and address the trauma that students experienced during the pandemic. We reaffirmed our commitment to prioritizing healing, care, and compassion in these conversations. We built stronger relationships with individual students and found new ways to focus on those with the most needs. The emphasis shifted from exam preparation to nurturing the well-being of our community, and there was a unanimous desire to sustain this focus.

Teachers also conveyed a collective feeling of satisfaction about our capacity to embrace digital tools and recognized the power of the many new platforms to expand learning outside of the classroom. Technology has been indispensable and necessary for our classrooms to become more inclusive in terms of student participation. It has also opened access to diverse perspectives, giving us the power to include voices and experiences from people of different backgrounds. Our lessons have become more personalized, and also multimodal, with videos, podcasts, infographics, and interactive simulations. Technology has also been critical for the advancement of independent learning. We saw evidence of this in weekly assignment posts, noticing some students completing all of their assignments well in advance. We also saw that some students were conducting independent research, as they included content in their assignments that was not provided by the curriculum.

Teachers shared that communication and collaboration between students had been fortified. The new opportunities to connect with peers online established a sense of community, even when we could not physically be together. Teachers expressed how the strength of relationships had encouraged students to feel more comfortable taking risks and relying more on each other for support. Teachers shared how, as a result of forming these relationships, more learners were able to significantly advance without the assistance of a teacher. In fact, we observed a clear increase in collaborative learning. For instance, we observed student-generated comments made on shared digital documents with peers. Students who were interacting in this collaborative manner outperformed their peers who did not engage collectively. Finally, at the end of two different units, 50 students were asked to rate their own academic performance using a collaboration self-assessment rubric. We then compared the culminating project scores for these students to their self-assessment data and found that those who rated themselves higher on the collaboration self-assessment rubric also scored higher on the project rubric. We agreed that we should continue planning instruction focusing on increasing opportunities for student collaboration.

It was powerful to me to see how so many of my colleagues recognized that we established a culture of trying new things. I remain deeply impressed by the high regard that teachers have expressed for the prevailing growth mindset within our community (Yeager & Dweck, 2020). It was particularly heartening to hear that veteran as well as novice educators were willing to experiment with new approaches during one of the most demanding periods of our professional lives. For instance, one teacher completely redesigned the math curriculum and converted every lesson into a digital interactive notebook. Another teacher started videorecording their lessons and converting these videos into interactive assessments. This influenced others, including me, to follow suit.

The most consequential experiment that persisted was the flipped classroom model that started for us at the beginning of the pandemic.

Insights about Challenges from Teacher Team Discussions: The Flipped Classroom

Throughout our teacher team discussions, we shared the many challenges we had in common as we lived through the seismic disruption the pandemic had on teaching and learning in our community. The greatest challenge that arose during our meetings was the decision to flip some of our classrooms. The flipped classroom model inverts the traditional teaching model, with idealized intentions of giving students responsibility for their own learning (Ozdamli & Asiksoy, 2016; Tang et al., 2023). For example, in a conventional flipped classroom model, teachers record direct instruction (e.g., lecture) to introduce or review the lesson's science content. Students then view these recordings as homework. Alternatively, students could prepare for class by reading text or carrying out another independent assignment. When they are in class, students work collaboratively on assignments, laboratory exercises, discussions, and projects. In this way, classroom time is dedicated to students working together rather than focusing on the teacher leading instruction. Thus, a consequential restructuring of the responsibilities of teachers must be part of this transition, as their role in the classroom shifts from lecture-based teaching to facilitating students' collaborative learning (Nouri, 2016).

Experimenting with Flipped Classrooms During Remote Teaching

I was one of the handful of educators in my school who experimented with this model. Every one of us found it challenging to keep students engaged and motivated in our flipped classrooms. We designed our assignments to be completed at any time, rather than in the classroom in real time, but many ended up not being completed. The intention of this model was to offer flexibility to accommodate our students. However, this inadvertently led to student challenges with time management, staying motivated, and remaining connected with peers and teachers. We heard students describe how they found it challenging to prioritize assignments, maintain focus, and attend support sessions. Without the accountability of the in-person classroom, students started to miss dead-lines, which led to poor academic performance, frustration, and anxiety. The absence of live interaction made it challenging for students to advance, and teachers had difficulty providing feedback and support, leading to students' feelings of frustration, disengagement, and dis-interest. Although flipping the classroom was a logical choice for us for many reasons, in hindsight, launching it as a beta version during a pandemic while students were dealing with countless other challenges may not have been the best move.

Efforts to Adjust the Flipped Classroom Model to Better Meet Our Students' Needs

We soon discovered that preparing for flipped classrooms required a hefty investment of time for planning and developing materials that was unsustainable, particularly when we were already struggling to cope with teaching during a pandemic. Additionally, with so many new online learning platforms, the digital realm kept expanding, adding to the strain. While we collectively believe that we did well, individually we struggled with this approach. Every teacher who attempted this strategy eventually became overwhelmed. Although some students thrived, many others struggled, leading us to question our approach altogether.

After considering the challenges faced in the flipped classrooms, we asked students for their feedback. Many students were challenged by this new routine and expressed apprehension about moving forward on their own. This was potentially due to a variety of reasons, including negative past experiences, personal or family issues, and learning differences or disabilities—each of which significantly contributed to their academic struggles. The lack of interaction and unequal access to technology led us to the conclusion that a fully flipped classroom was not appropriate for our community of learners.

Weekly Checklist

These checklists were inserted right at the start of our Google Slides presentations, each of which was structured to outline learning objectives and tasks that the students were expected to accomplish within a span of a week. Having observed student challenges with focus and prioritization, there were multiple rationales for our decision to implement a weekly assignment checklist at the end of the 2019–2020 school year during remote teaching (see Figure 9.2). First, the checklist facilitated organization and planning, breaking down the tasks into manageable portions to help students stay on top of their workload and prioritize tasks based on deadlines and difficulty levels. Second, the interactivity of the checklist promoted students' monitoring their own progress, thereby fostering a sense of responsibility and independence. Third, by coupling assignments with relevant vocabulary, the checklist integrated language learning with content learning. Finally, the digital nature of the checklist allowed for quick updates and adjustments, providing a dynamic tool that can be adapted for each week's learning objectives or based on student and teacher feedback. This element of adaptability made the checklist a living document that evolved with the learning process. We hoped the checklist would support our students in increasing their academic independence, while still providing a safety net of structure and guidance. This approach combined the benefits of structure with self-guided learning, and appeared to embody our teacher team's commitment to fostering a dynamic, student-centered learning environment.

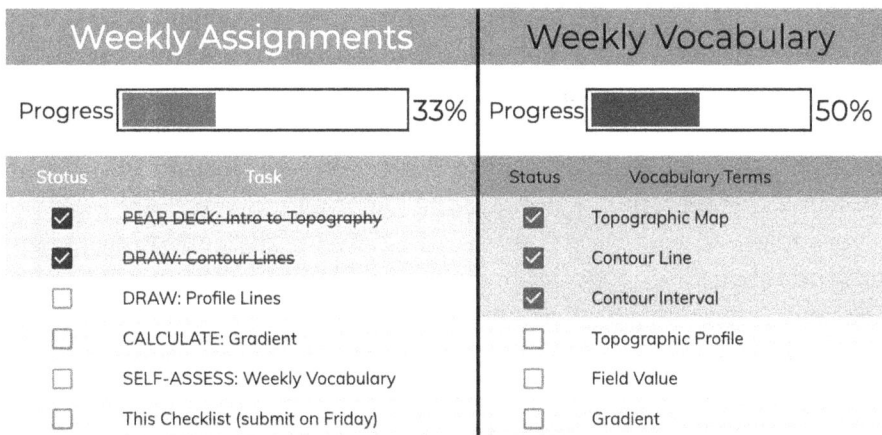

Weekly Assignments		Weekly Vocabulary	
Progress	33%	Progress	50%
Status	**Task**	**Status**	**Vocabulary Terms**
☑	PEAR DECK: Intro to Topography	☑	Topographic Map
☑	DRAW: Contour Lines	☑	Contour Line
☐	DRAW: Profile Lines	☑	Contour Interval
☐	CALCULATE: Gradient	☐	Topographic Profile
☐	SELF-ASSESS: Weekly Vocabulary	☐	Field Value
☐	This Checklist (submit on Friday)	☐	Gradient

Figure 9.2 A weekly assignment and vocabulary checklist showing progress towards completion as a student works towards the end-of-week self-assessment activity. Screenshot taken by the author, and checklist developed by the author.

What Did We Learn from Remote Teaching Using a Flipped Classroom during the Pandemic?

We learned a lot from our work together, which we owe to the collaborative and supportive teacher team culture that evolved. We believe that our school community-oriented approach drew on our diverse backgrounds and experiences. The inclusive and collaborative teacher-learning environment kept our spirits high and allowed us to keep high expectations for ourselves as well as for our students.

The pandemic provided an opportunity for us to revamp our teaching practices and create a more equitable system that attends to more students' needs. We flipped our classes in an attempt to do this, but I believe we were not successful at first. Although our initial attempt fell short of our expectations, I felt the flipped model could be modified further, because we learned so much from our shortcomings, and my teacher team agreed.

One insight that our teacher team gleaned from our experiences is the importance of continually seeking student feedback on the weekly interactive checklist. Students' input can play a crucial role in fine-tuning the tool to meet their individual learning needs and preferences. As part of this feedback process, we offered students a comprehensive list of terms related to their coursework and asked them to identify those they deemed most important or challenging. This provided valuable insights to us regarding our students' perceptions and understandings of the course material. Moreover, we believed that incorporating students' feedback into the planning and refinement of the interactive checklist supported a more inclusive and equitable learning environment where they could feel that their voices were heard and valued.

Through our experiences, we also learned about the numerous benefits of recorded lessons. These recorded lessons provided students with flexibility to learn at their own pace. However, we've come to appreciate the potential for peer interaction around the interactive weekly assignment and vocabulary checklist. As students worked independently at varying paces and stages of completion, they could support one another in navigating assignments and strategies.

I view the checklist as essential to our fully flipped classroom approach. We first introduced it during remote teaching at the end of the 2019–2020 school year. Over time, the checklists underwent modifications and became interactive when the majority of staff began using Pear Deck, an extension of the Google Slides platform, during the 2020–2021 academic year. This technological enhancement enabled students to make annotations directly on the slides and retain a copy of it for personal use. An advantage of this approach was its minimal demand for alterations to the original presentation slides. Using Pear Deck, we explored and tested multiple versions and variations of the weekly checklist. All these explorations were driven by the common goal of assisting students in managing and organizing their tasks, especially since virtual interactions with them were limited to once or twice per week.

Heading Back to the Classroom: Takeaways That Informed the Modified Flipped Classroom Approach

It became apparent through our remote teaching that not all students enjoyed equal access to resources and technical support outside of school. To combat this disparity and promote equitable learning opportunities, we decided to assign asynchronous assignments to be completed in school rather than at home. We felt that this recognized and respected the different circumstances students faced at home, ensuring that all learners had access to the resources and environment they needed to successfully complete their work. To help students become more successful with our modified strategy moving forward, we learned that we needed to provide more initial structure, gradually introducing freedom and choices while offering a more complete range of support along the way.

With the return to in-person instruction, some educators gradually deviated from using the weekly checklists. However, there remained an interest in the idea of checklists within each department. With the belief that refining our checklist systems would be beneficial for both our fellow teachers and students, we began to explore alternative platforms. In the 2021–2022 academic year, we started to experiment with Google Sheets to implement checklists.

Modifying a Flipped Classroom Model in My Earth Science Classroom

A flipped classroom involves students watching videos or other online resources before coming to class, and then using class time for activities that reinforce and apply the material they have already seen. A modified in-class flipped classroom presents students with opportunities to collaborate on assignments and offers increased face-time to work with the teacher individually or in small groups. Traditional flipped classrooms largely depend on internet use outside of school, an aspect that can be challenging for some students due to technical difficulties or lack of internet access. The inclusion of in-class sessions can mitigate these issues by providing students with access to technology and the internet within the school premises, thereby reducing inequities. Furthermore, this model allows for immediate teacher feedback and support, a crucial factor in aiding students who may find the material challenging. Overall, I felt that this modified in-class flipped classroom model had potential to offer balanced and inclusive teaching, encouraging student ownership and fostering student choice.

Upon our return to in-person teaching, during the 2021–2022 academic year, in an effort to create a more student-centered classroom based on what we learned from the flipped classroom experiment, I opted to conduct a trial implementing a modified in-class flipped approach. To start, I divided my students into small learning groups of four. I began with a full week of lessons that focused on a common task, standard, or lab activity.

Modification #1: Fun Facts for Learning about Each Other
Each class period began with a segment called "Fun Facts, News, & Announcements," designed to foster interpersonal knowledge. Incorporating this five-minute community-building activity at the start of each class seemed to have an impact. The elimination of time spent on giving instructions (because they were on the checklist) allowed us to share experiences prior to delving into the science content, maximizing instructional time while fostering relationship-building. A few minutes dedicated to listening to students' fun facts and news seemed to contribute to a sense of community and strengthen interpersonal relationships. It offered students an opportunity to understand each other's interests, perspectives, and experiences, thereby promoting an inclusive and supportive learning atmosphere. This daily routine also encouraged the development of communication skills and empathy.

Modification #2: Framing and Assessments
In addition to the "Fun Facts, News, & Announcements," I framed each lesson by addressing the questions of "What are we doing? Why are we doing it? How do I know I am doing a good job?" I developed short assessments for my students to take immediately after engaging with the content, and longer assessments to demonstrate a deeper understanding. I merged all related slides into one presentation for the students to go through on their own. Additionally, I chose short practice activities for

students to complete and recorded all direct instruction for the week, keeping each video to five minutes or less and embedded checks for understanding. Throughout the week, I planned small group instruction as students worked asynchronously, keeping a journal of students' progress towards learning goals.

Modification #3: Adapting the Weekly Interactive Checklist and Small Group Work

I used the weekly checklist of all the tasks students should complete and presented it to the whole class with a multiple-choice agenda (see Figure 9.3). I modified the interactive checklist to provide a suggested sequence in which tasks could be completed, while maintaining that students choose their daily tasks. I thought that offering a suggested order as a scaffold would provide structured guidance and possibly aid students with time management. I gave students a choice on what tasks they would do each day and targeted specific students for one-on-one and small group instruction. Each week ended with a full class discussion, self-assessment, and learning reflection before repeating the process.

Modification #4: More Small Group Work and Time with Students

Since we started every class with "Fun Facts" to help build community, I no longer had to spend time giving instructions since they were embedded in

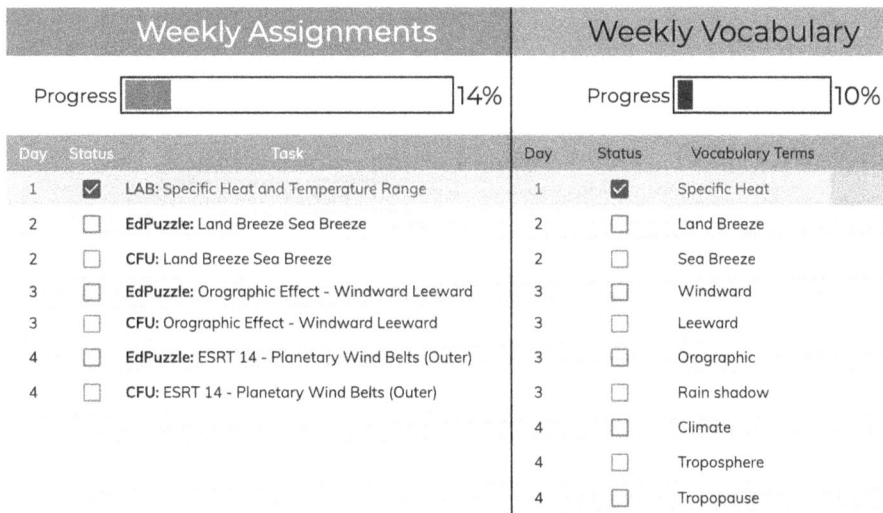

Weekly Assignments			Weekly Vocabulary		
Progress	14%		Progress	10%	
Day	Status	Task	Day	Status	Vocabulary Terms
1	☑	LAB: Specific Heat and Temperature Range	1	☑	Specific Heat
2	☐	EdPuzzle: Land Breeze Sea Breeze	2	☐	Land Breeze
2	☐	CFU: Land Breeze Sea Breeze	2	☐	Sea Breeze
3	☐	EdPuzzle: Orographic Effect - Windward Leeward	3	☐	Windward
3	☐	CFU: Orographic Effect - Windward Leeward	3	☐	Leeward
4	☐	EdPuzzle: ESRT 14 - Planetary Wind Belts (Outer)	3	☐	Orographic
4	☐	CFU: ESRT 14 - Planetary Wind Belts (Outer)	3	☐	Rain shadow
			4	☐	Climate
			4	☐	Troposphere
			4	☐	Tropopause

Figure 9.3 An example of a weekly checklist of tasks developed for the modified flipped classroom model. Screenshot taken by the author, and checklist developed by the author.

the weekly checklist. With this additional time, I was able to devote more attention to small group work. Additionally, I have continued to revise and use pre-recorded instruction and interactive slides made during the pandemic. Students access and process these resources independently, freeing me to allocate more time to small group instruction and offer tailored guidance and personalized attention to individuals. The increased frequency of small group instruction helps me understand my students better, both in terms of their academic strengths, struggles, progress, and who they are as individuals. This has been a tremendous resource that helps inform my planning and instruction.

In the modified flipped classroom, I can spend more time with small groups of students. I have more opportunities for feedback and dialogue, which helps me build trust and rapport with my students. I feel this has allowed our class to be more inclusive and I think a stronger supportive learning environment. I am also better able to incorporate relevant materials and activities, leading to a more engaging and meaningful learning experience. For example, one of my students had been facing challenges in learning English. With more time for small group instruction, I worked with this student and provided extended support. This allowed me to build a stronger relationship with the student, which in turn appeared to allow the student to be more engaged in the classroom. As a teacher, I feel that being able to work closely with individual students through small group instruction helped me to create a more culturally responsive and sustaining classroom environment.

Students' Experiences

What follows is my understanding of students' experiences with a modified flipped classroom. I learned through informal conversations with students that being in our science class is special and enjoyable for most. One aspect of this class that stands out to the students is the opportunity to work in groups and collaborate with peers for the majority of the time. Unlike other classes where students tend to work individually, they expressed that our science class provides an environment where they can discuss their ideas more frequently, and work with their friends to complete assignments. While students still have fun joking and gossiping, they are also asking each other questions and completing their tasks in a relatively productive manner.

Any teacher reading this might wonder how the freedom provided does not cascade into chaos. Indeed, the initial phase was a bit chaotic, but we eventually found our rhythm and I believe we have been able to be productive, in part, by redefining productivity to include having fun and making connections with each other. Over time, we have nurtured strong relationships within the class. A sense of camaraderie is rooted in trust, enjoyment, and student autonomy, starting from something as fundamental as choosing their own seats. This desire must be respected (as well as leveraged), and I have promised that students will only move seats (of their own volition) after we decide together what would be best for them moving forward. So far this has worked. The students consistently expressed that learning alongside their friends constituted the most enjoyable aspect of their school experience. Students explained that our class lets them get to know their teacher and classmates on a more personal level.

A distinguishing feature of our science class was our daily incorporation of technology into lessons and activities, thus fostering digital literacy skills development. Unlike other classes with one assignment per day, in our class, students have a list of different tasks to complete which they can choose to tackle in a way that works for them. Because of this, students have the flexibility to tackle these tasks in a manner that suits them best.

Weekly Checklist

Implementing this ambitious design demands detailed organization, so everything students are accountable for is entered and shared on an interactive, digital checklist. The intention is to help students track their own progress and help them stay aware of what they need to accomplish by each checkpoint. The length between checkpoints starts small to give students practice with tracking their work before extending to cover two to three days, gradually building in size. Our checklists include all tasks, their associated vocabulary, the resources that contain the vocabulary, and a suggested pacing guide. At the end of each week, students are guided through a reflection based on the checklist that includes a space for student feedback. Because I believe that students should have a say on how their education is administered, I provide this space for them to share their experiences and to make suggestions. It is through these reflections that I have learned how to tweak the checklist format. Students consistently expressed their support for how we are now running our classes. They shared little to no negative feedback and confirmed that the choice and flexibility provided helps them feel engaged, empowered, and more in control of their learning.

Closing Thoughts

Teachers communicate a vast amount of information, both overtly and subtly, through their actions, decisions, and demeanor. Students are perceptive and pick up on everything, including the expectations that teachers convey through the space they give for student voice and freedom. These expectations can become a self-fulfilling prophecy, and a rigid schedule can limit independent engagement and compliance may make life easier for the teacher, but it does not benefit the student (Jackson & Zmuda, 2014). It is better to nurture defiance, encouraging students to challenge, question, make choices, and if necessary, stand in opposition. By allowing and encouraging students to make decisions about their own learning, teachers communicate trust and can guide defiance into independence. By encouraging students to make independent decisions, teachers contribute to their cognitive and moral growth. Giving students the independence to steer their learning path communicates the teacher's faith in their capabilities.

Reflecting on my own practice, I believe that modifying a conventional flipped classroom model to an in-class flipped model allowed me to provide a more balanced and inclusive approach to learning that has potential to benefit all students. This approach allows for flexibility and provides opportunities for students to learn together, hopefully building a stronger sense of community. Moreover, by ensuring that all students have physical access to the necessary materials and resources in the classroom, it can reduce the potential of inequities. With this approach, I can provide immediate feedback and support, and develop a learning environment where my students can work at their own pace in class.

CRSE is an approach that requires continuous learning and adaptability, and my colleagues and I have learned a lot from our work together. Through collaboration, we created a supportive community-oriented environment that met the diverse needs of our students. We have also learned the benefits of asynchronous learning in conjunction with differentiated instruction to meet individual student needs. However, we also recognized the need for providing initial structure and support, as well as incorporating student feedback to promote inclusivity.

In conclusion, our experiences with CRSE, and a modified in-class flipped classroom model, have taught us the importance of flexibility, inclusivity, and collaboration. We have gained a deeper appreciation for the power of diverse perspectives and experiences and recognize the need for a tailored approach to instruction to meet individual students' needs. Through ongoing learning and growth, we hope to continue to improve our teaching practices and provide a more inclusive and equitable learning environment for all our students. We hope you do, too.

Reflecting on the World on Fire

I felt like we had been teaching while the world around us burned to the ground, and I think it is still on fire. I remember reading, "We can't change this world or put out these fires unless we engage and activate the minds and hearts of ourselves and our students" (Delpit, 2019, p. 12). Delpit presented this imagery to emphasize the importance of teaching with a social justice lens, recognizing that, as educators, we have a responsibility to empower our students to become agents of change in their own lives, communities, and in the world. Her imagery has seared itself into my mind and I find the fervid flames a fitting phrase; however, I also see the fire as a hopeful analogy. For the better part of a century, naive humans suppressed wildfires at all costs until we looked closer. We now understand the natural cycle of wildfire includes transition and rejuvenation. In cases where equilibrium is significantly disrupted, the forest transitions and restores itself without reverting to its previous state. In fact, it often establishes a biome that is even more robust than its prior condition. Nevertheless, the development of the succeeding community hinges on what happens in the immediate aftermath.

This fiery imagery has indelibly imprinted itself in my mind. While it indeed presents a vivid picture of urgency, I also interpret it as a beacon of hope. For many years, humans misunderstood the role of wildfires, striving to extinguish them without fully comprehending their purpose in the natural order. Now we appreciate that wildfires play an essential role in the cycle of transition and rejuvenation. When the equilibrium is severely disrupted, the forest doesn't merely revert to its previous state; instead, it evolves and regenerates, often into a version even more resilient than before. This metaphor extends beyond the natural world, offering crucial lessons for our current situation. The future of the community that will emerge from the current disruption depends significantly on our immediate actions and responses. Thus, this imagery serves as a potent reminder of the need for swift and thoughtful actions (Delpit, 2019, p. 12).

Reflection and Discussion Questions

Inspired by the shock of remote learning, I found myself delving into the extensive literature on education reform because of a deep desire to actively engage in the ongoing discourse surrounding the necessary changes in post-pandemic teaching. As you, the reader, peruse these words, I sincerely hope that you will contribute your own thoughts and perspectives, fostering a collective conversation that extends beyond these pages.

1   What were some of the pedagogical decisions that you made during the switch to remote learning? Were there any approaches that you decided to keep even after you returned to the classroom in person? If so, why?
2   What are some of the approaches that you use in your classroom to promote your students' learning independently? Are there ways in which you believe that this connects to CRSE?
3   What are some ways that you seek to foster an inclusive environment in your classroom?

References

Bali, M. (2015, March 30). *Embracing subjectivity*. Hybrid Pedagogy. https://hybridpedagogy.org/embracing-subjectivity/

Chowning, J., Osuga, H., Bryant, W., & Foster, J. (2022). *Attending to race and identity in science instruction (Practice Brief #89)*. University of Washington Institution for Science + Math Education. https://stemteachingtools.org/brief/89

Delpit, L. (Ed.). (2019). *Teaching when the world is on fire*. The New Press.

Gay, G. (2010). *Culturally responsive teaching: Theory, research, and practice* (2nd ed.). Teachers College Press.

Godsil, R. D. (2016, May 23). Why race matters in physics class. *UCLA Law Review Discourse*, 64, 40–63. https://uclalawreview.org/wp-content/uploads/2019/09/Godsil-D64.pdf

Haberman, M. (2010). The pedagogy of poverty versus good teaching. *Phi Delta Kappan*, 92(2), 81–87. 10.1177/003172171009200223 (Reprinted from "The pedagogy of poverty versus good teaching," 1991, *Phi Delta Kappan*, 73[4], 290–294, http://ed618.pbworks.com/f/Habermann,+M.+Pedagogy+of+Poverty.pdf)

Hammond, Z. (2014). *Culturally responsive teaching and the brain: Promoting authentic engagement and rigor among culturally and linguistically diverse students*. Corwin Press.

Hammond, Z. (2021). Liberatory education: Integrating the science of learning and culturally responsive practice. *American Educator, 45*(2), 4–11. www.aft.com/summer2021/hammond

Hofstra, B., Kulkarni, V. V., Munoz-Najar Galvez, S., He, B., Jurafsky, D., & McFarland, D. A. (2020). The diversity-innovation paradox in science. *Proceedings of the National Academy of Sciences, 117*(17) 9284–9291. 10.1073/pnas.1915378117

Jackson, R., & Zmuda, A. (2014). Four (secret) keys to student engagement. *Educational Leadership, 72*(1), 18–24. https://www.newenglandssc.org/wp-content/uploads/2015/11/Jackson_Zmuda_4-keys-to-student-engagement.pdf

Ladson-Billings, G. (1995). Toward a theory of culturally relevant pedagogy. *American Educational Research Journal, 32*(3), 465–491. 10.3102/00028312032003465

López, F. A. (2017). Altering the trajectory of the self-fulfilling prophecy: Asset-based pedagogy and classroom dynamics. *Journal of Teacher Education, 68*(2), 193–212. 10.1177/002248711 6685751

Love, B. L. (2020). Teachers, we cannot go back to the way things were. *Education Week, 29.* www.edweek.org/leadership/opinion-teachers-we-cannot-go-back-to-the-way-things-were/2020/04

Milner, H. R. (2021). *Start where you are, but don't stay there: Understanding diversity, opportunity gaps, and teaching in today's classrooms.* Harvard Education Press.

New York City Department of Education (NYC DOE). (n.d.). *School quality snapshot 2021–2022.* Retrieved December 20, 2023, from https://tools.nycenet.edu/snapshot/2022

New York State Education Department (NYSED). (2019). *Culturally responsive-sustaining education framework.* http://www.nysed.gov/crs/framework

Nouri, J. (2016). The flipped classroom: for active, effective and increased learning – especially for low achievers. *International Journal of Educational Technology in Higher Education, 13,* Article 33. 10.1186/s41239-016-0032-z

Ozdamli, F., & Asiksoy, G. (2016). Flipped classroom approach. *World Journal on Educational Technology: Current Issues, 8*(2), 98–105.

Tang, T., Abuhmaid, A. M., Olaimat, M., Oudat, D. M., Aldhaeebi, M., & Bamanger, E. (2023). Efficiency of flipped classroom with online-based teaching under COVID-19. *Interactive Learning Environments, 31*(2), 1077–1088. 10.1080/10494820.2020.1817761

Yang, S., Tian, H., Sun, L., & Yu, X. (2019). From one-size-fits-all teaching to adaptive learning: The crisis and solution of education in the era of AI. *Journal of Physics Conference Series, 1237*(4), 042039. 10.1088/1742-6596/1237/4/042039

Yeager, D. S., & Dweck, C. S. (2020). What can be learned from growth mindset controversies? *American Psychologist, 75*(9), 1269.

Part

4

Collaborative Chapters

Stories and Reflections

10

When the World Tilted Differently

Exploring Science Teachers' Pandemic Stories through a Culturally Responsive and Sustaining Science Lens

Jamie Wallace, Elaine V. Howes, Maya Pincus, Kin Tsoi, Raghida Nweiran, Susan Bullock Sylvester, Arthur W. Funk, Caity Tully Monahan, Samantha Swift, and Sean Krepski

---

HIGHLIGHTS

- This chapter examines the CRSE PLG's collaborative inquiry into teachers' reflections on their stories told during the early days of the COVID-19 pandemic. We grounded our exploration in the questions: *What do teachers notice when reflecting on their stories during the pandemic? How do teachers engage in sensemaking of their teaching and learning experiences during a time of disruption and transformation?*
- Three themes emerged in our analysis of our reflections on our pandemic stories: *Centering students' care, well-being, and community; losing human connection during remote teaching,* and *teacher support and learning during the intertwined pandemics.*
- Our collaborative chapter illustrates a metacognitive approach, as we reflected on and analyzed stories of our teaching and thinking that we had previously shared together.
- We hope you take away from this chapter inspiration for engaging in storytelling with colleagues about experiences during these pandemics, as well as possibilities for supporting students in discussing important and challenging issues in science using a critical stance.

DOI: 10.4324/9781003397977-15

Introduction

In this chapter, we collaboratively explore teachers' reflections on the stories that the members of our professional learning group told during the early days of the pandemic. Throughout the second year of our work together, the Culturally Responsive and Sustaining Education Professional Learning Group (CRSE PLG) had been meeting regularly to study CRSE, share and analyze teaching experiences, and develop culturally responsive instructional strategies. And then the COVID-19 pandemic hit our city …

Through the heartbreaking effects of the pandemic, vast health, racial, and socioeconomic inequities, along with the systemic racism that causes these, were made starkly visible. Given this context, this chapter focuses primarily on our analysis of our teacher stories during the beginning of the COVID-19 pandemic in New York City (NYC). "Stories from the Field" is a conversational routine at each meeting, where we share observations, thoughts, and insights about our classrooms, instruction, and educational settings in the context of CRSE. This routine provides grounding in our own everyday examples, allowing us to provide insights into our individual school contexts and reflect on our teaching practice. Thus, for this exploratory study, we wondered, *What do teachers notice when reflecting on the stories told during the pandemic? How do teachers engage in sensemaking of their teaching and learning experiences during a time of disruption and transformation?*

As discussed in Chapter 1, the CRSE PLG uses four tenets of CRSE to inform our work. The group originally started with four fundamental tenets of culturally relevant education (Howes & Wallace, 2022; Wallace et al., 2022b); we have since expanded them to encompass culturally responsive and sustaining science teaching. Our CRSE tenets entail teaching that is asset-based, using students' assets as resources for teaching and learning (López, 2017; NYSED, 2019); draws connections to students' cultures to strengthen and sustain them (Ladson-Billings, 1995a, 2014; Paris & Alim, 2017); holds high expectations for all students' academic learning (Gay, 2010; Ladson-Billings, 1995b); and adopts and supports students in developing a critical stance toward sociopolitical structures and processes (Ladson-Billings, 1995b; Paris & Alim, 2017). These tenets provide a frame for this study as we explore our reflections on our stories during the first months of the pandemic.

Why Stories?

Storytelling is a universal narrative tradition featured across cultures and a method for learning about how humans experience and make sense of the world (Connelly & Clandinin, 1990). As anthropologist Mary Catherine Bateson wrote, "The human species thinks with metaphors and learns through stories" (1994, p. 11). In teaching and teacher education, storytelling provides pictures of our individual school environments and allows us to reflect and process experiences from our professional lives. As teacher researchers studying our own practice, stories provide a tangible way to offer insights into our own identities as science teachers in relation to others (Beauchamp & Thomas, 2009).

In education, storytelling is a method that can tap into teachers' "personal professional knowledge" (Clandinin, 1999/2015; Clandinin & Connelly, 2004), and a way to bring teacher learning and expertise into public discussions. Focusing on distinct situations, teachers often share their thinking and knowledge of teaching through stories, drawing on their own experiences. This kind of storytelling is built into the cultures of teaching (Feiman-Nemser & Floden, 1984). Storytelling is instrumental not only as a tool for teacher learning but also as a means for working toward change and transformation. Storytelling has also been prevalent in anti-racist pedagogy (Alderman et al., 2021) and social justice advocacy (Novak, 2021). In the CRSE PLG, we have found storytelling to be a rich, descriptive mechanism for us to engage in sensemaking and explore culturally responsive education together (Wallace et al., 2022b).

With our explicit focus on CRSE and learning more about the students in our classrooms and their lived experiences (Moll et al., 1992), the CRSE PLG uses Stories from the Field to share and puzzle through instructional practices and classroom situations (Wallace et al., 2022b). We recognize there is an aspect of curation of the stories we choose to share in this space, allowing us to construct and reconstruct moments and memories from the past and present, and contemplate the future (Connelly & Clandinin, 1990). Teachers' stories are often situational and contextual, and can range from a specific problem or challenge, to a puzzlement, exemplar, or scenario. Stories on practice are typically told after the situation has occurred, often with hopes of communicating a particular message (Lampert, 1999), or receiving advice on a challenge. Stories are living records that awaken with each retelling, allowing the narrator to relive meaningful moments while listeners accompany them on the journey (Clandinin, 1999/2015). Aspects of stories may be removed, constructed, or fabricated with space for numerous interpretations, thus portraying mosaics with multiple dimensions rather than mirrors on personal experiences (Lampert, 1999). Instead of mirrors, teachers' stories can serve as windows into their practice and their own perceptions of their practice, whereby one can catch glimpses and fragments of rooms in the house but not the house in its entirety.

Told in the context of professional communities, listeners often feel compelled or have a responsibility to respond. Listener responses may take the form of advice, interpretation, critique, or even adding on with sharing similar experiences. As Clandinin (2015) reflects, teachers do not often have an opportunity or designated space to reflect on their relationships with the students in their classrooms. In the context of the CRSE PLG, stories provide opportunities for teachers to examine their own and fellow science teachers' thinking and experiences regarding CRSE in a safe and familiar environment.

In our routine, storytelling became a collaborative process where our experiences often intertwined as related to one another, sharing part of ourselves. Stories had a particular flow and group members asked questions inquiring into practice, spurring reflection, clarifying details and specifics, and often providing suggestions or alternative options. In this way, teachers gave each other support in the practice of CRSE, embracing stories as a pedagogical tool to enhance their learning and understanding of themselves, their teaching, and their students (Coulter et al., 2007; Marshall, 2023; Wallace et al., 2022b). CRSE Stories from the Field developed into a process for sensemaking and collaboratively interpreting stories from a teacher-researcher perspective, to think critically about the situation or interaction from a place of inquiry, and question the nature of our own thinking and the sociopolitical structures at play, while also considering the perspectives of the actors involved (Kincheloe, 2000; 2012; Kincheloe et al., 2011; Marshall, 2023).

In studying CRSE, we have learned that developing relationships and learning about one's students are essential. A routine of storytelling can provide a structure and space in which to reflect on the ways in which teachers and students interact with each other. Stories provide a vehicle for teachers to reveal the challenges and celebrate the successes that they experience in building relationships with their students. These stories form a narrative collage that recounts the endeavors of teachers working to create more culturally responsive classroom environments.

Context: Intersecting Pandemics of COVID-19 and Racial Injustices

In this section, we draw on multiple data sources, including examples that teachers in the group used in their schools, to provide background on how our teaching intersected with the pandemic and informed our collaborative work during this time. We use teaching artifacts to offer insights into our context and lived experiences during the time that the stories were told, providing snapshots from work the teachers used in their classrooms and schools.

In late August 2019, the CRSE PLG resumed regular biweekly meetings to begin our second year of work together. In December 2019, we began hearing about COVID-19 in Wuhan, China. In mid-January, countries outside of China, including the United States, confirmed pandemic-related deaths; the first confirmed case in NYC was on February 29, 2020 (NYC Health, n.d.; Taylor, 2021). On March 13th, the American Museum of Natural History (our home institution) along with multiple cultural institutions in NYC closed their doors (Pogrebin & Cooper, 2020; Wallace et al., 2021). Mayor Bill de Blasio ordered schools to close on March 16th, when NYC became an epicenter of the global pandemic (Knoll & Sedacca, 2020; McKinley, 2020). Within this rapid time line, life changed in traumatic and turbulent ways as the death rates surged and we were thrust into quarantine with an uncertain ending and path forward.

Early in the pandemic, meeting together as a PLG became easier in some ways, as we came together remotely. It also became more necessary, as we viewed these meetings as a way to stay in touch, to figure out together how to teach remotely and how CRSE could still take place in this strange new world. Leading professional learning about CRSE for practicing NYC science teachers through the American Museum of Natural History's Gottesman Center for Science Teaching and Learning became another component of the group's work together. Thus the structure of our time and work together shifted to weekly online meetings.

We had begun engaging in teacher research in earnest in the fall of 2019; in fact, we had just presented on the initial stages of our inquiries at the annual Ethnography Forum at the University of Pennsylvania in February 2020 a month before the COVID-19 pandemic broke out on the East Coast, with our city as its epicenter. Therefore, some of the studies started before the pandemic, while others were products of it; in all cases, the pandemic played a role in our thinking and understanding of CRSE in science.

As of July 2023, NYC had lost 45,321 of its inhabitants to COVID-19 (City of New York, 2023). The students in our schools are from communities hardest hit by this disease. During the pandemic's most challenging days, before the advent of vaccinations, our schools were shut down and teaching went remote. The COVID-19 pandemic shone a light on those who did and did not have access to health care, housing, food, electronic devices, and digital technologies and infrastructures that were needed to function during lockdown and quarantine. Early in the pandemic, influential voices in culturally responsive education and critical race theory called for a "hard re-set." They urged us to use the crisis to reconceive what teaching and learning could look like—with CRSE and anti-racist pedagogy as grounding pillars—to make sure that we were centering and reaching our children and setting them up for academic, social, and cultural success (Ladson-Billings, 2021a; Love, 2020). Discussions in academic circles not only shifted to online environments but also to reenvisioning how things could and needed to look different going forward.

In addition to highlighting the inequalities that were made exceedingly apparent, and calls for action regarding social justice, some of the research on teaching and education early in the pandemic focused on socioemotional learning (DeArmond et al., 2021; Hamilton et al., 2021), humanizing pedagogies (Blum & Dale, 2021; Bozkurt & Sharma, 2021; Carter Andrews et al., 2019; Chand et al., 2022; Freeman et al., 2020; Mehta & Aguilera, 2020), teacher well-being and mental health (Dabrowski, 2020; Kush et al., 2022), pedagogies of care (Buckley-Marudas & Rose, 2021), transitions to remote teaching and learning (Wallace et al., 2021), and maintaining relationships with students (Carter Andrews et al., 2021).

The challenges of remote teaching ran parallel to teachers' desire to engage with their students about the widely publicized instances of police violence directed at Black people and the rise of the Black Lives Matter movement in the summer of 2020. The CRSE PLG believed that attending to their students and providing safe spaces for them to talk during this time was fundamental to their work as culturally responsive and sustaining science teachers. To help illustrate what this looked like in our individual classrooms, we share samples of teacher artifacts accompanied with brief commentary (see Figures 10.1–10.4).

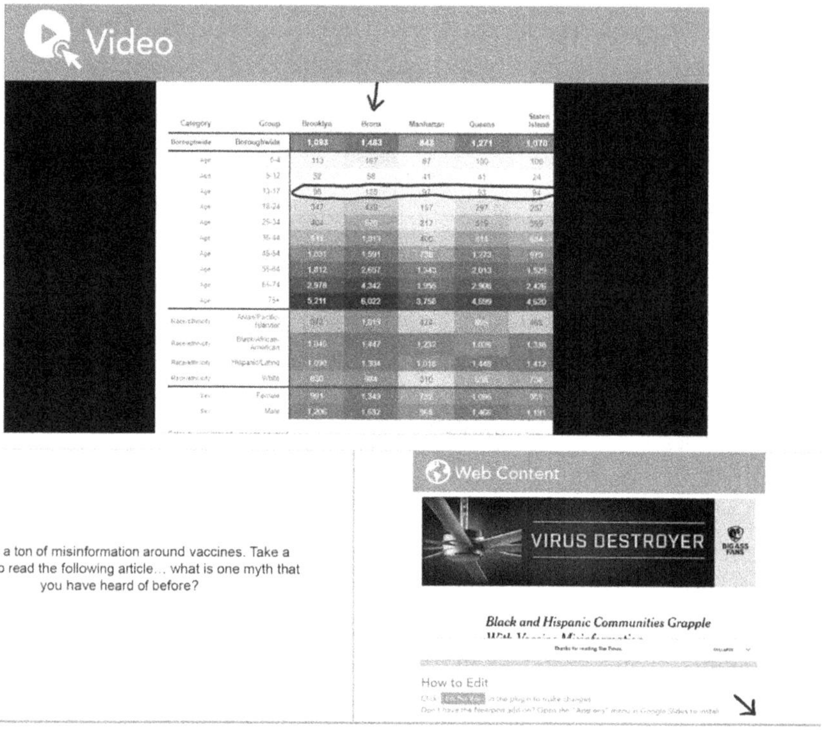

Figure 10.1 Slides from a NearPod that Raghida Nweiran developed and shared in class highlighting the number of COVID-19 cases in the school's area affecting students' ages, misinformation, and how the virus is unproportionately affecting communities. Screenshots and images developed by Raghida Nweiran.

Raghida: These slides did not connect to content I was teaching at the time. I thought the information was important enough that students needed to see it, because it affected their communities. Science does not fit into discrete boxes, so even though my students were only enrolled in Earth science at the time, I saw the benefit of pausing our usual curriculum for this. We connected it to the process of science and how representation matters, for things like vaccine creation and understanding how medicines may affect bodies differently. We also spoke about how historically communities of Color were medical guinea pigs.

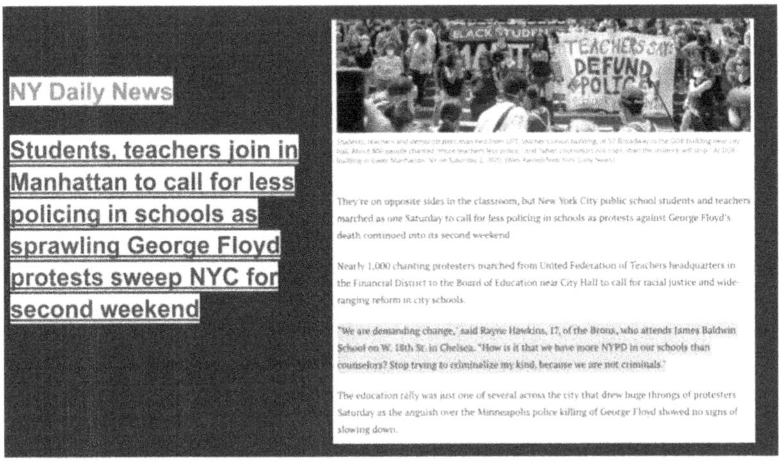

Figure 10.2 Slide from Caity Tully Monahan used during a crew/advisory meeting on June 11, 2020, featuring an article from the *NY Daily News* calling for reduced policing in schools following George Floyd protests. Slide developed by Caity Tully Monahan.

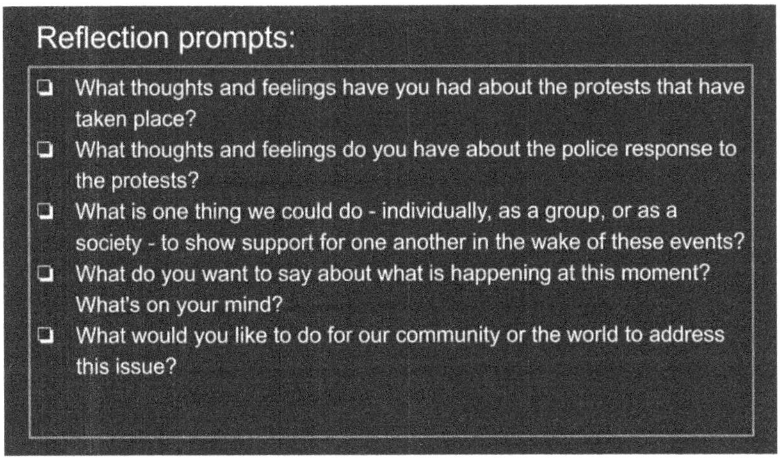

Figure 10.3 A slide that Susan Bullock Sylvester used in a town hall at her school on June 8, 2020, featuring reflection prompts to consider following an increase in Black Lives Matter protests. Slide courtesy of Susan Bullock Sylvester.

Susan: This slide was part of a school-level approach to "critical caring" through virtual town hall meetings. This was a collaboration of five people on the town hall committee that included the school's dean, social worker, advocate counselor, and two teachers.

During this time, racial, social, and economic disparities were brought to the fore and we saw how certain communities and neighborhoods were hit harder by COVID-19 and racial violence. The death of George Floyd on May 25, 2020, ignited protests and marches frequently led by the Black Lives Matter Movement (see Figure 10.4 for a time line). A call to action in hopes of making the world stop and confront the radical injustices perpetrated on marginalized communities, and Black communities in particular, was prevalent in public discourse.

The timing of this study overlapped with the pandemic and recent racial justice movements; thus, these are inextricably linked with our findings and learning from this collaborative inquiry process. The COVID-19 pandemic halted individual teacher research studies. The contemporaneous advocacy protests spurred reflection on our ongoing inquiry from perspectives originating in racial justice movements. The collaborative aspect of the group was heightened during this time as the group chose to meet more frequently, sharing stories about interactions with students and lessons taught related to racial justice movements.

Positionality

Our exploration of our positionality is central to our work together and informs our lived experiences, interpretations, and teaching. Our teachers identify our group as majority white or white presenting, nearly evenly split in cisgender males and females, and unified in our endeavor to better serve our students. We recognize that we do not represent the majority of our students in terms of race, ethnicity, or socioeconomic upbringing; this has indeed been a focus in many of our conversations, and became even more prominent during the multiple pandemics. Each teacher is employed in a school in NYC that serves a majority of students classified as economically disadvantaged. As Merriam and colleagues argued, "the reconstruing of insider/outsider status in terms of one's positionality vis-a-vis race, class, gender, culture and other factors, offer us better tools for understanding the dynamics of researching within and across one's culture" (2001, p. 405). It is a journey that we continue to examine and navigate, and one that we feel is important to disclose in the context of our stories, interpretations, and our

Murder Time Line

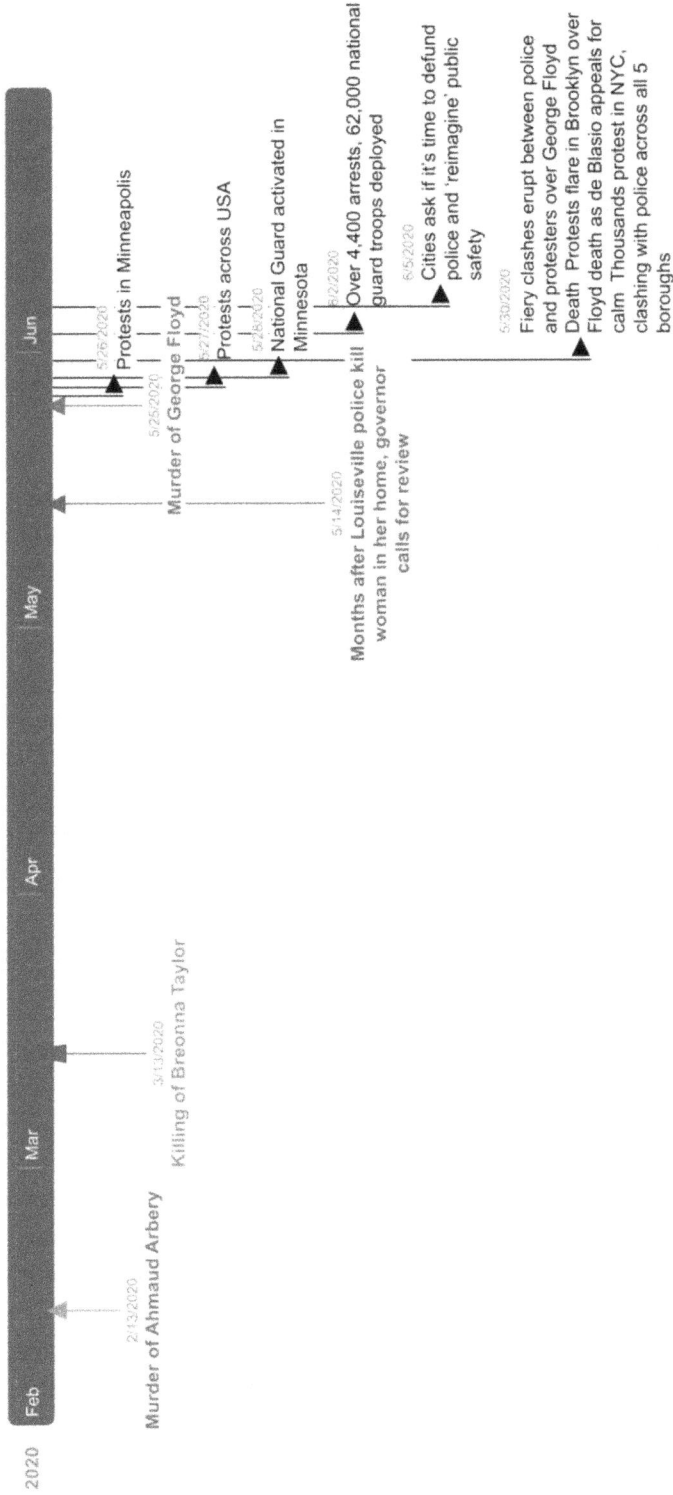

Figure 10.4 This Murder Time Line provides a chronology of the reported deaths, related protests, and events from February through June 2020, starting with the murder of Ahmaud Arbery. The time line helps to depict what was happening across the country at this time. Image courtesy of Wallace et al., 2022a.

teacher research. Part of this journey is examining our own identities and biases, and how they provide a lens on how we perceive the world. We agree with scholars' arguments (Fenge et al., 2019; Milner, 2007) that recognizing our racial and cultural positionality is a necessary component of research in order to critically consider, confront, or avoid potential dangers that may surface in research and in our teaching.

Culture is very much contextual, and informs our interpretations and experiences. We know that our positionality informs our interpretations, sensemaking, and search of meaning. We look to anthropology to examine ways in which culture has been defined and described previously. Anthropology forefather Clifford Geertz referenced part of Kluckhohn's twenty-seven page definition of the concept of culture with eleven aspects including "the total way of life of a people" and "a way of thinking, feeling, and believing'" (1973, p. 311). Geertz continued on to provide his own thoughts on the concept of culture, stating "Believing, with Max Weber, that man is an animal suspended in webs of significance he himself has spun, I take culture to be those webs, and the analysis of it to be therefore not an experimental science in search of law but an interpretive one in search of meaning" (p. 311). Geertz later described how the study of culture is connected to analysis of social discourse noting, "In the study of culture the signifiers are not symptoms or clusters of symptoms, but symbolic acts or clusters of symbolic acts, and the aim is not therapy but the analysis of social discourse" (1973, p. 320). We find this quote particularly fitting as this chapter focuses on an analysis of teachers' reflections on their CRSE stories through discourse (e.g., storytelling).

Methods

In this collaborative study, we used qualitative methods as participant observers and teacher researchers engaging in inquiry. We wondered what themes we would see in the stories that we told throughout the year, and how stories would provide a window into our relationships with our students during this time. We used a collaborative research design (Christianakis, 2010) informed by culturally responsive research practices (Berryman et al., 2013; Trainor & Bal, 2014). The collaborative nature of this study provided an opportunity for reciprocal learning among group members "to help one another reach a greater understanding of their pedagogy and, thus, better serve their respective students" (Christianakis, 2010 p. 116). CRSE PLG teachers selected the stories of focus for the study and collaboratively engaged in analyzing their own stories, and provided feedback, together with and alongside facilitators. The facilitators documented and transcribed interviews containing stories, worked with teachers in the analysis, and led the writing of the first draft.

We collected multiple sources of data and bounded the time frame of the study to investigate the second year of the group in particular, looking at stories that were told between September 2019 to July 2020 (Creswell, 1998; Miles & Huberman, 1994). Data sources include CRSE PLG meetings transcripts, from which the facilitators copied each teacher's stories throughout the year, and presented them chronologically and individually, so that each teacher could review the stories they told. We also used teachers' graphic organizers and discussion notes as data. With 40 meetings throughout the 2019–2020 year, the group convened consistently over an estimated 80 hours. In our analysis, we took a metacognitive approach to review and reflect on transcripts of our individual stories (e.g., Meeting on 1.14.21) and then discussed what we noticed together. We also incorporated our current reflections and reflective memos on our stories as another data source and analytical tool. Where possible, we drew on data sources and artifacts that teachers in the group used in their schools to talk about the pandemic and/or racial justice movements, including NearPod, Google Slides, and websites. We also incorporated time lines on the pandemic and racial justice movements, drawing on current events as reported in the *New York Times*, to situate our work within the larger context that was plaguing our and our students' lived experiences.

Analysis

In an earlier paper exploring teachers' stories from the first year of the group, we engaged in coding before our analysis (Wallace et al, 2022b). Given the pandemic climate and stress, we changed our methods to a more flexible approach to allow for reflections, current thinking, and to add another layer onto the stories. We adopted a reflective approach to our collaborative analysis, using prompts to consider when looking at the data. Teachers reflected on excerpts of their own stories from the transcripts compiled by the facilitators. To assist teachers in analyzing their own stories, facilitators also provided a "See, Think, Wonder Note Catcher" (hereafter "STW")—a scaffold that teachers frequently use in their own classrooms (see Figure 10.5 for a modified STW template from Project Zero, 2019). For our purposes, we embedded the following prompts within the tool: "*What do you see in your stories? Which of your stories from during the pandemic are resonating with you now and why? What do you notice in relation to culturally responsive*

See, Think, Wonder Note Catcher

We are using this familiar routine to think about and reflect on data from Stories from the Field during the start of the pandemic. It is meant to offer structure and support when looking at the data in making observations and interpretations. We also provide a few prompts to consider when looking at the data.

(S) See	(T) Think	(W) Wonder
What do you **see** in your stories?	What do you **think** about what you see?	What does it make you **wonder**?
Which of your stories from during the pandemic are resonating with you now and why?	*What are you thinking now?*	*Do the stories raise any questions for you?*
What do you notice in relation to culturally responsive education?		

Figure 10.5 Scaffold for thinking about and analyzing pandemic stories data called a See, Think, Wonder Note Catcher. Image developed by Wallace and Howes (Wallace et al., 2022a), modified from Project Zero, 2019.

*education? What do you **think** about what you see? What are you thinking now? What does it make you **wonder**? Do the stories raise any questions for you?"*

We decided on this approach for several reasons: 1) the tool was a familiar scaffold to support teachers' analyses of their stories as it was something they had used in their own classrooms, 2) it supported organization of thinking and metacognition over brief durations of time, and 3) it provided structure and an element of distance from data that was personal and emotional in nature about a particularly challenging time.

Teachers completed the tool as they read through their stories transcripts to capture their thinking and reflections. They then discussed the following prompts in pairs before engaging in a group discussion: *What do you notice? Are there any similarities or threads emerging across your analyses? Which stories did you find particularly intriguing, and why? [Is there] Anything else that you find important, including questions?* These reflections, initial thinking memos, and notes from discussions helped us to examine our research questions. In the course of two remote meetings, teachers further shared and commented upon their individual and small group notes together as a whole group. These individual, small group, and whole group reflections, taken together, supported our collaborative analytical approach.

Our analytic approach was a metacognitive exercise grounded in teachers' thinking about thinking, using familiar teaching tools. Studies suggest that more research is needed on teachers' thinking, writing, and discussion about their thinking (Wilson & Bai, 2010) and that professional learning communities are an environment that can nurture and support teachers' metacognitive skills with potential for impacting the learning of others (Prytula, 2012). Although teachers are increasingly expected to teach students metacognitive skills, there is limited research on *teachers'* awareness and exploration of their own metacognition and pedagogical understanding of metacognition, illuminating a need in teacher education (Wilson & Bai, 2010). Research also suggests that metacognition can play an important and powerful role in teacher professional learning and in influencing classroom instruction (Jiang et al., 2016), and resonates with teacher research as it supports reflection on teaching (Duffy et al., 2009).

Reflecting on Our Pandemic Stories

Our analysis of our Stories from the Field suggests three interrelated themes concerning our experiences in the early months of the pandemic: *Centering caring for students*, *coping with loss of social interaction during remote teaching*, and *teacher support and learning during the intertwined pandemics.* Next, we elaborate on each of the themes.

Centering Students' Care, Well-Being, and Community

It became clear through our analysis that teachers' stories forefronted students' socioemotional well-being during the early months of the pandemic. The CRSE PLG members consistently noted that this need took precedence over academic demands, in individual classrooms, in schools, and in the wider educational community. While teacher–student relationships had always been a theme in the group's discussions about CRSE, it became a larger focus as these relationships became more tenuous and harder to maintain during remote teaching.

Kin summed up this shift from academics to supporting students' well-being in his reflection noting, "Critical caring seems to come into play at some point here. The idea that I felt the need to more check-in on them and make sure they're okay and to see if there's anything that I could do to help them with their struggles." Reflecting on his stories, he commented, "I just kind of stopped teaching [content] after a while. I literally say [in one story], 'I stopped teaching content; I don't care about it right now; I'd rather make sure the kids are okay.'"

Susan reflected on a story she told on her school's approach to critical caring. She shared how her school held virtual town hall meetings, attended by school personnel and students at least once a week (see Figure 10.3). The town hall meetings, often led by teachers, and sometimes featuring guest speakers, addressed issues related to both the pandemic and the racial justice movement. Students were encouraged to participate in these meetings even if they could not make it to class. As Susan reflected, the town halls provided "an important opportunity for students to engage in a way that they did not feel put on the spot academically." In this way, Susan described how her school valued caring and social connection among students and school personnel, and treated it as a priority, temporarily setting aside typical academic pressures on students in order to support them through an unprecedented crisis.

In her STW, Caity described how she was communicating with her students' parents, and supporting them, indicating an extension beyond a vision of a school community that includes only students and teachers. She wrote:

> When calling parents today, it was clear that they were at their wits end and just listening to them and assuring them that they're doing as good of a job as they can be. … It was reassuring to them that they're not a bad parent because their kid didn't hand in an assignment or that they got a call from me and a social worker.

Throughout these reflections on stories, we also see an emphasis on community coming through in different dimensions. Susan highlighted the town halls that her school held to foster school community. Caity talked about interacting with her students' parents, another component of the school community. Kin focused on the community in his classroom, and centering his students' well-being.

Losing Human Connection during Remote Teaching

In March of 2020, schools shut down in NYC, and instruction was suspended for a couple of weeks during quarantine. At this time, teachers worked with colleagues and school administrators to decide what their remote instruction would look like. While teachers worked to develop approaches for remote instruction that could engage students in science, and support them emotionally and socially, the enforced physical distancing from students was an ongoing worry and sorrow.

Missing in-person, classroom-based, social interaction arose several times in the STW stories reflections. The constraints of remote teaching reminded Susan that she felt better able to support her students' learning and interactions when in person:

> I see that to meet the students where they were I had to scale back the content—which reminded me that [when] we are present in the classroom [we can] ask those higher level DOK [Depth of Knowledge] questions and get the students talking with each other. The isolation was so difficult—I had a few students who seemed to prefer remote learning while others just seemed to check out.

Helping students talk with one another was something that did not come easily with remote teaching, especially early in the pandemic. Susan's reflections also highlighted the challenging nature of feeling isolated, and that there were differences in how students were responding to the situation. For instance, she noted that some students seemed to prefer remote learning while others "just seemed to check out," either by not attending or by not engaging with the lessons. Caity's reflection points out the equity-related concern that not all students had access to a reliable internet connection, and she noted in her STW that in "late May people start losing access to the Internet." Without access to Wi-Fi connections, the "check-out" phenomenon would intensify, as students literally were unable to "check in."

In their reflections on their stories, teachers noted feeling disconnected from students in the remote environment. In addition to highlighting that many students did not have access to a functioning Internet connection, Caity lamented that online teaching at that time had felt "dehumanizing and robotic." Reflecting on one of her stories during the early days of remote teaching she shared:

> [I] felt a little bit like I'm not doing my job because I feel so much like a robot. And I'm feeling guilty that I feel like I'm not doing my job and like what else should I be doing? So it's all these emotions, and usually there's such an emotional lift to our jobs—I feel a lot of it's just cut off right now.

Others also reflected on this robotic feeling expressed in their stories, and this became especially apparent when reviewing stories from before the pandemic. Raghida strongly emphasized this idea, reflecting:

> My stories in 2020, right before lockdown, were all like "the kids said this," "and we dressed up like our cultures," and it's so sweet and

lovey and I'm sharing all of my jokes. And then my two stories after we went remote were like, "I'm a robot. I posted [in] Google classroom. I have no idea what's happening in the kids' lives." I just detached, so quickly, and lost the human connection.

Going beyond the loss of everyday interaction and relationships with students, Kin noted that "I wish I'd had a chance to say goodbye to some of those kids. Now just reading [my stories transcripts] it's like … 'I never got to say bye because they never showed up for the Zoom meeting.'" Sam (who had spent the spring of 2020 teaching remotely in Los Angeles (LA)) said that reviewing her stories

> brought me back to LA [where] the last chunk of the time I was there was just teaching virtually … . I'll never see those students again. … I continuously talk about how it started off, we had a good chunk of kids coming, and then they slowly got off. But there were even some kids that never showed up once online. It was just like as soon as we went virtual, [I] never got to hear from them again.

This theme does not have a happy ending. But it brings to the fore the importance of the relationships that teachers build with students through their day-to-day interactions. The loss of these interactions was keenly felt and surfaced in the reflections as sadness and loss. It is as if the teachers felt they became de-humanized as part of the technology, rather than as human beings interacting with other living breathing human beings.

Teacher Support and Learning during the Intertwined Pandemics: Sharing Approaches and Reflection

Teachers and schools had a couple of weeks between the time when schools shut down for quarantine and their reopening to plan for remote instruction. Within our group of teachers and their respective schools, multiple types of schedules, virtual teaching platforms, and expectations for instruction were selected and developed.

Updates and awareness of individual and school approaches. In the CRSE PLG, teachers used Stories from the Field to share their individual and school plans for remote instruction with each other. For instance, Sean's stories depicted how he, with other teachers at his school, collaboratively revised his Earth science curriculum to make it more cohesive and student-centered (see Chapter 4). Arthur used the months of remote teaching to develop his goal to "not go back to the way things were," (Love, 2020) reflecting, "I think I benefited more from this change than a lot of people. It just kind of shook me in a way" (see Chapter 9). In reflecting on her pandemic stories, Susan referred to "on-the-fly problem solving" as the pandemic "was unfolding moment by moment, and some of those challenges were on the individual level: technologies, student engagement, isolation, loss of a sense of time … And wondering and worrying about our students. How do we help them, in the midst of this?"

An emphasis on reflection and support. In the course of analyzing their stories, the importance of the CRSE PLG as a supportive community for the teachers, including having a space for thinking, processing, and discussion, arose several times. As Susan reflected,

> We're very fortunate to have this group because teaching needs to be reflective and talking about it among a group that understands what we're going through is very helpful. And I think that is what happened during the pandemic: We're all going through the same time line, but experiencing it slightly differently with different student populations.

Pandemic stories were told during the health crisis and racially driven events we were experiencing during February through June of 2020 (see Figure 10.4). Reflection on stories told during this time brought back the intense, unprecedented situations group members were describing during the early days of the pandemic. Caity summarized this when reflecting on the group's stories in general: "We're talking about COVID, talking about parents dying, talking about Ahmaud Arbery, and then … George Floyd. And then it's every week, it's something big. And it's like holy sh--. What a year."

Arthur described the importance of having a space for refuge and support, noting that the CRSE PLG meetings provided

> the place that has helped me the most, just talking these things through. So much of my professional time is spent with kids or trying to help another teacher do their job. Being able to be in a place where we all have this thing in common, that we're trying to do this together, and everybody here has something to contribute … There isn't another place in my life where I get to explore these sorts of thoughts.

Susan reflected similar ideas, with

> I think for me it was that the collaboration of this professional learning community was really powerful in that we could share ideas. That we could express what we were going through and then share resources. …

I noticed [that] I can't get over how, in the midst of the pandemic, we had just the horror of the Ahmaud Arbery lynching, we had George Floyd murdered, we had awareness that the pandemic was hitting communities that were already marginalized.

Discussion

Teachers' stories shared during the early days of the pandemic provided an outlet for support and a place for struggle during a devastating and challenging time. Teachers often do not have a designated space for reflection and opportunity to consider their relationships and interactions with their students (Clandinin, 1999/2015). Within the CRSE PLG, teachers' reflections on their pandemic stories offered a chance for teachers to process what they and their students were experiencing both in the moment and in the recent past. Additionally, teachers' analysis of their own pandemic stories within the group provided opportunities to engage in metacognitive thinking within a supportive environment (Prytula, 2012; Wilson & Bai, 2010).

Teachers' STW scaffolds and discussions centered largely around relational issues, based in communities: classroom communities, school communities, and the larger national community. The lens of remote teaching influenced all discussions of pedagogy. Social disjunctures engendered by the pandemic permeated our talk and our work together during this time. The feeling of loss centered around missing in-person interactions with students; some went a whole academic year without meeting students face to face. The everyday, taken-for-granted aspects of teaching grounded in relationships with students and with school colleagues were gone. This feeling heightened our awareness of these vitally important relationships and their role in our CRSE-centered teaching.

Our analysis and reflection on our pandemic stories also brought to the fore an emphasis on critical care and prioritizing the well-being of our students. This connects to Tyrone Howard's (2021) integration of care as one of five key principles of culturally responsive pedagogy as "an authentic and culturally informed notion of care for students, wherein their academic, social, emotional, psychological, and cultural well-being is adhered to" (p. 140). This notion of care was solidified for teachers, especially during the pandemic, and is evident in their stories with a commitment to the well-being of their students.

In connecting with literature on CRSE during this time as a group, we found the following quote from Gloria Ladson-Billings particularly powerful as it resonated with our work, especially around the theme of centering students' care and well-being:

> The hard re-set demanded by the COVID-19 and anti-Black pandemics of 2020 require us to engage in culturally relevant pedagogy that takes into account the conditions of students' lives these occurrences set in motion. Specifically, we must re-set around technology, curriculum, pedagogy, assessment, and parent/community engagement that will support and promote students' culture. (2021a, p. 73)

The turmoil of the multiple pandemics made more visible during this time impacted teachers across the world as they needed to learn to teach remotely, engage and care for students in different ways, and keep themselves healthy and sane. Indeed, since that time, scholars like Ladson-Billings have identified four pandemics simultaneously occurring: COVID-19, systemic racism, environmental catastrophe, and looming economic crisis (Ladson-Billings, 2021b). In this context, analyzing and reflecting on our pandemic stories provided insights about a time that offered little sense, and served to clarify our thinking about CRSE in uncertain times. Additionally, as an educational and cultural support, they allowed us to breathe and re-set (Ladson-Billings, 2021a), "navigate pain and possibility" (Lyiscott, 2019, p. 89), confront educational inequities, and continue centering CRSE and our students' experiences in our work.

In terms of methodology, using a scaffold that the teachers frequently integrate into their own teaching was a helpful entry to engage in metacognitive thinking around stories within the context of our group, a skill that is supportive of critical reflection on teaching with potential for powerful implications for influencing classroom instruction (Duffy et al., 2009; Jiang et al., 2016; Wilson & Bai, 2010). While teachers were very interested in analyzing their stories from early in the pandemic and eager to engage in the work, it was important for us to approach it sensitively and cautiously, knowing how painful and traumatic the time was for each of us and our communities. We found that using a familiar scaffold was one way to approach the analysis systematically and offer prompts to support reflection and conversation, providing a tool that could be responsive to our group and our needs at that time. In some ways, our analysis of our stories functioned as an additional layer of processing the events and experiences we had recently lived through. Disseminating this analysis through a teacher researcher lens appeared to us as a needed and somewhat untapped area in teacher education (see Figure 10.6).

Figure 10.6 Photograph of some of the members of the CRSE PLG presenting a collaborative poster about our analysis of pandemic stories at the 2022 American Educational Research Association (AERA) annual meeting in San Diego, California. Photograph taken by a participant.

Implications

This study has helped to clarify our thinking around CRSE, especially in the current sociopolitical context. Our findings about what we noticed in our pandemic stories reinforce what we see in the literature and push our thinking to consider the importance of adding a dimension of humanizing pedagogy, care, and well-being into our CRSE tenets (Carter Andrews et al., 2019; Howard, 2021).

Engaging in collaborative research and professional learning, this study explored how teachers managed the early days of the intertwined pandemics. Through this collaborative work, the teachers supported one another and shared accounts of and insights into their ongoing lived experiences. At this point, the pandemic lingers and remote teaching in K-12 schools is no longer an everyday occurrence for most. There will always be a need for teacher adaptation and change, and a nonjudgmental space for critical self-reflection (Dodo Seriki, 2018) and learning to provide collegial understanding and support through unforeseen circumstances.

References

Alderman, D., Perez, R. N., Eaves, L. E., Klein, P. & Muñoz, S. (2021). Reflections on operationalizing an anti-racism pedagogy: Teaching as regional storytelling. *Journal of Geography in Higher Education*, 45(2), 186–200. 10.1080/03098265.2019.1661367

Bateson, M. C. (1994). *Peripheral visions: Learning along the way*. HarperCollins Publishers.

Beauchamp, C., & Thomas, L. (2009). Understanding teacher identity: An overview of issues in the literature and implications for teacher education. *Cambridge Journal of Education*, 39(2), 175–189. 10.1080/03057640902902252

Berryman, M., SooHoo, S., Nevin, A., Arani Barrett, T., Ford, T., Joy Nodelman, D., Valenzuela, N., & Wilson, A. (2013). Culturally responsive methodologies at work in education settings. *International Journal for Researcher Development*, 4(2), 102–116. 10.1108/IJRD-08-2013-0014

Blum, G. I., & Dale, L. (2021). Becoming humanizing educators during inhumane times: Valuing compassion and care above productivity and performance. *Current Issues in Education*, 22(3), 1–17. 10.14507/cie.vol22iss3.1992

Bozkurt, A., & Sharma, R. C. (2021). On the verge of a new renaissance: Care and empathy oriented, human-centered pandemic pedagogy. *Asian Journal of Distance Education*, 16(1), i-vii. http://asianjde.com/ojs/index.php/AsianJDE/article/view/576

Buckley-Marudas, M. F., & Rose, S. E. (2021). Collaboration, risk, and pedagogies of care: Looking to a postpandemic future. *The Journal of Interactive Technology & Pedagogy*, 19. https://jitp.commons.gc.cuny.edu/collaboration-risk-and-pedagogies-of-care-looking-to-a-postpandemic-future/

Carter Andrews, D. J., Brown, T., Castillo, B. M., Jackson, D., & Vellanki, V. (2019). Beyond damage-centered teacher education: Humanizing pedagogy for teacher educators and preservice teachers. *Teachers College Record*, 121(6), 1–28. 10.1177/016146811912100605

Carter Andrews, D. J., Richmond, G., & Marciano, J. E. (2021). The teacher support imperative: Teacher education and the pedagogy of connection. *Journal of Teacher Education*, 72(3), 267–270. 10.1177/00224871211005950

Chand, R., Alasa, V., & Chand, R. D. (2022). "Humanizing" pedagogies in online learning and teaching-A necessity in the wake of the Covid-19 pandemic. *Journal of Positive School Psychology*, 6(10), 3713–3722. http://mail.journalppw.com/index.php/jpsp/article/view/13915

Christianakis, M. (2010). Collaborative research and teacher education. *Issues in Teacher Education*, 19(2), 109–125. https://www.itejournal.org/wp-content/pdfs-issues/fall-2010/14christianakis.pdf

City of New York. (2023). *COVID-19: Data*. https://www.nyc.gov/site/doh/covid/covid-19-data-totals.page.

Clandinin, D. J., & Connelly, F. M. (2004). *Narrative inquiry: Experience and story in qualitative research*. John Wiley & Sons.

Clandinin, D. J. (2015). Stories to live by on the professional knowledge landscape. *Waikato Journal of Education, 20*(3), 183–193. 10.15663/wje.v20i3.233 (Reprinted from "Stories to live by on the professional knowledge landscape," 1999, *Waikato Journal of Education, 5,* 107–120. 10.15663/wje.v5i0.403)

Connelly, F. M., & Clandinin, D. J. (1990). Stories of experience and narrative inquiry. *Educational Researcher, 19*(5), 2–14. 10.3102/0013189X01900500z

Coulter, C., Michael, C., & Poynor, L. (2007). Storytelling as pedagogy: An unexpected outcome of narrative inquiry. *Curriculum Inquiry, 37*(2), 103–122. 10.1111/j.1467-873X.2007.00375.x

Creswell, J. W. (1998). *Choosing among five traditions: Qualitative inquiry and research design.* Sage Publications, Inc.

Dabrowski, A. (2020). Teacher wellbeing during a pandemic: Surviving or thriving? *Social Education Research, 2*(1), 35–40. 10.37256/ser.212021588

DeArmond, M., Chu, L., & Gundapaneni, P. (2021). *How are school districts addressing student social-emotional needs during the pandemic?* Center on Reinventing Public Education. https://crpe.org/how-are-school-districts-addressing-student-social-emotional-needs-during-the-pandemic/

Dodo Seriki, V. (2018). Advancing alternate tools: Why science education needs CRP and CRT. *Cultural Studies of Science Education, 13,* 93–100. 10.1007/s11422-016-9775-z

Duffy, G. G., Miller, S., Parsons, S., & Meloth, M. (2009). Teachers as metacognitive professionals. In D. J. Hacker, J. Dunlosky, & A. C. Graesser (Eds.), *Handbook of metacognition in education* (pp. 240–256). 10.4324/9780203876428

Feiman-Nemser, S., & Floden, R. (1984). *The cultures of teaching* (Occasional Paper No. 74). Michigan State University. Institute for Research on Teaching, College of Education, Michigan State University. https://citeseerx.ist.psu.edu/document?repid=rep1&type=pdf&doi=c9b7425e5adaafb6695b513c52a073b539d4a36b

Fenge, L. A., Oakley, L., Taylor, B., & Beer, S. (2019). The impact of sensitive research on the researcher: Preparedness and positionality. *International Journal of Qualitative Methods, 18,* 1–8. 10.1177/1609406919893161

Freeman, Q., Flores, R., Garzón, D., Gumina, D., Sambolín Morales, A. N., Silva Diaz, E., & Stamatis, K. M. (2020). Collaborating towards humanizing pedagogies: Culture circles in teacher educator preparation. *The New Educator, 16*(1), 86–100. 10.1177/1609406919893161

Gay, G. (2010). *Culturally responsive teaching: Theory, research, and practice* (2nd ed.). Teachers College Press.

Geertz, C. (1973). Thick description: Toward an interpretive theory of culture. In *The interpretation of cultures: Selected essays* (pp. 3–30). Basic Books, Inc. https://books.google.com/books?id=BZ1BmKEHti0C&pg=PA3&source=gbs_toc_r&cad=2#v=onepage&q&f=false

Hamilton, L., Gross, B., Adams, D., Bradshaw, C. P., Cantor, P., Gurwitch, R., Jagers, R., Murry, V. M., & Wong, M. (2021). *How has the pandemic affected students' social-emotional well-being? A review of the evidence to date [Research report].* Center on Reinventing Public Education. https://crpe.org/how-has-the-pandemic-affected-students-social-emotional-well-being-a-review-of-the-evidence-to-date/

Howard, T. C. (2021). Culturally responsive pedagogy. In J. A. Banks (Ed.), *Transforming multicultural education policy & practice: Expanding educational opportunity* (pp. 137–163). Teachers College Press.

Howes, E., & Wallace, J. (2022, August 9). Exploring culturally responsive science teaching through turbulence and challenge: Starting a multi-year research study during the pandemic. *AAAS Advancing Research & Innovation in the STEM Education of Preservice Teachers in High-Need School Districts (ARISE).*

Jiang, Y., Ma, L., & Gao, L. (2016). Assessing teachers' metacognition in teaching: The teacher metacognition inventory. *Teaching and Teacher Education, 59,* 403–413. 10.1016/j.tate.2016.07.014

Kincheloe, J. (2012). *Teachers as researchers (classic edition): Qualitative inquiry as a path to empowerment.* Routledge. 10.4324/9780203801550

Kincheloe, J. L. (2000). Making critical thinking critical. *Counterpoints, 110,* 23–40.

Kincheloe, J. L., McLaren, P., & Steinberg, S. R. (2011). Critical pedagogy and qualitative research. In Norman K. Denzin & Yvonna S. Lincoln (Eds.), *The SAGE handbook of qualitative research* (4th ed.) (pp. 163–177). Sage Publications.

Knoll, C., & Sedacca, M. (2020, March 16). Parents scramble as N.Y.C. schools close over coronavirus. *The New York Times.* https://www.nytimes.com/2020/03/16/nyregion/nyc-schools-closed-coronavirus.html?smid=url-share

Kush, J. M., Badillo-Goicoechea, E., Musci, R. J., & Stuart, E. A. (2022). Teachers' mental health during the COVID-19 pandemic. *Educational Researcher, 51*(9), 593–597. 10.3102/0013189X221134281

Ladson-Billings, G. (1995a). Toward a theory of culturally relevant pedagogy. *American Educational Research Journal, 32*(3), 465–491. 10.3102/00028312032003465

Ladson-Billings, G. (1995b). But that's just good teaching! The case for culturally relevant pedagogy. *Theory Into Practice, 34*(3), 159–165. 10.1080/00405849509543675

Ladson-Billings, G. (2014). Culturally relevant pedagogy 2.0: aka the remix. *Harvard Educational Review, 84*(1), 74–84. 10.17763/haer.84.1.p2rj131485484751

Ladson-Billings, G. (2021a). I'm here for the hard re-set: Post pandemic pedagogy to preserve our culture. *Equity & Excellence in Education, 54*(1), 68–78. 10.37256/ser.212021588

Ladson-Billings, G. (2021b). Three decades of culturally relevant, responsive, & sustaining pedagogy: What lies ahead? *The Educational Forum, 85*(4), 351–354.

Lampert, M. (1999). Knowing teaching from the inside out: Implications of inquiry in practice for teacher education. *Teachers College Record, 100*(5), 167–184. 10.1177/016146819910000507

López, F. A. (2017). Altering the trajectory of the self-fulfilling prophecy: Asset-based pedagogy and classroom dynamics. *Journal of Teacher Education, 68*(2), 193–212. 10.1177/0022487116685751

Love, B. L. (2020). Teachers, we cannot go back to the way things were. *Education Week, 29.* www.edweek.org/leadership/opinion-teachers-we-cannot-go-back-to-the-way-things-were/2020/04

Lyiscott, J. (2019). *Black appetite. White food.: Issues of race, voice, and justice within and beyond the classroom*. Routledge.

Marshall, S. L. (2023). In pieces: An approach to critical reflexivity with science teachers through storytelling and in community. *Journal of Science Teacher Education, 34*(5), 563–581. 10.1080/1046560X.2023.2169304

McKinley, J. (2020, March 22). New York City region is now an epicenter of the coronavirus pandemic. *The New York Times*. https://www.nytimes.com/2020/03/22/nyregion/Coronavirus-new-York-epicenter.html?searchResultPosition=9

Mehta, R., & Aguilera, E. (2020). A critical approach to humanizing pedagogies in online teaching and learning. *International Journal of Information and Learning Technology, 37*(3), 109–120. 10.1108/ijilt-10-2019-0099

Merriam, S. B., Johnson-Bailey, J., Lee, M. Y., Kee, Y., Ntseane, G., & Muhame, M. (2001). Power and positionality: Negotiating insider/outsider status within and across cultures. *International Journal of Lifelong Education, 20*, 405–416.

Miles, M. B., & Huberman, A. M. (1994). *Qualitative data analysis: An expanded sourcebook* (2nd ed.). Sage Publications, Inc.

Milner, H. R. (2007). Race, culture, and researcher positionality: Working through dangers seen, unseen, and unforeseen. *Educational Researcher, 36*(7), 388–400. 10.3102/00131 89X07309471

Moll, L. C., Amanti, C., Neff, D., & Gonzalez, N. (1992). Funds of knowledge for teaching: Using a qualitative approach to connect homes and classrooms. *Theory Into Practice, 31*(2), 132–141. 10.1080/00405849209543534

New York City (NYC) Health. (n.d.). *COVID-19: Data*. City of New York. Retrieved September 2, 2021, from https://www1.nyc.gov/site/doh/covid/covid-19-data-trends.page#outcomes

New York State Education Department (NYSED). (2019). *Culturally responsive-sustaining education framework*. http://www.nysed.gov/crs/framework

Novak, L. (2021). Review of *Storytelling for social justice: Connecting narrative and the arts in antiracist teaching* [Review of the book *Storytelling for social justice: Connecting narrative and the arts in antiracist teaching*, by L. A. Bell]. *Studies in Art Education, 62*(2), 195–199. 10.1080/00393541.2021.1896267

Paris, D., & Alim, H. S. (Eds.). (2017). *Culturally sustaining pedagogies: Teaching and learning for justice in a changing world*. Teachers College Press.

Pogrebin, R., & Cooper, M. (2020, March 12). New York's major cultural institutions close in response to coronavirus. *The New York Times*. https://www.nytimes.com/2020/03/12/arts/design/met-museum-opera-carnegie-hall-close-coronavirus.html

Project Zero. (2019). See, think, wonder. Harvard Graduate School of Education. http://www.pz.harvard.edu/sites/default/files/See%20Think%20Wonder_2.pdf

Prytula, M. P. (2012). Teacher metacognition within the professional learning community. *International Education Studies, 5*(4), 112–121. 10.5539/ies.v5n4p112

Taylor, D. B. (2021, March 17). A timeline of the coronavirus pandemic. *The New York Times.* https://www.nytimes.com/article/coronavirus-timeline.html

Trainor, A. A., & Bal, A. (2014). Development and preliminary analysis of a rubric for culturally responsive research. *The Journal of Special Education, 47*(4), 203–216. 10.1177/0022466912436397

Wallace, J., Howes, E. V., Funk, A., Krepski, S., Pincus, M., Sharif, R., Swift, S., Tsoi, K., Sylvester, S., & Tully, C. (2022a, April 21–26). *When the world tilted differently: Science teachers' pandemic stories through culturally responsive-sustaining education* [Poster presentation]. American Educational Research Association Annual Meeting, San Diego, CA, United States. 10.3102/IP.22.1893926

Wallace, J., Howes, E. V., Funk, A., Krepski, S., Pincus, M., Sylvester, S., Tsoi, K., Tully, C., Sharif, R., & Swift, S. (2022b). Stories that teachers tell: Exploring culturally responsive science teaching. *Education Sciences, 12*(6), 401. 10.3390/educsci12060401

Wallace, J., MacPherson, A., Hammerness, K., Chavez-Reilly, M., & Gupta, P. (2021). Pivoting in a pandemic: Supporting STEM teachers' learning through online professional learning during the museum closure. *Journal of STEM Outreach 4*(3). 10.15695/jstem/v4i3.11

Wilson, N. S., & Bai, H. (2010). The relationships and impact of teachers' metacognitive knowledge and pedagogical understandings of metacognition. *Metacognition and Learning, 5*(3), 269–288. 10.1007/s11409-010-9062-4

11

Reflections

Learning in, from, and with the CRSE PLG

*Kin Tsoi, Maya Pincus, Susan Bullock Sylvester,
Raghida Nweiran, Sean Krepski, Samantha Swift,
Arthur W. Funk, Elaine V. Howes, and Jamie Wallace*

HIGHLIGHTS

- In this chapter, we reflect on why we formed the CRSE PLG and what we learned from participating.
- Considering some of the challenges encountered in learning to engage in teacher research, we offer insights into what we would have liked to have known about CRSE as beginning teachers.
- To close the chapter, we offer "A Gift": A document of instructional resources for science teaching including sociopolitical situations, websites, documentaries, activist groups, and stories.
- We hope that our reflections on our time together will resonate with you and inspire you to consider participating in long-term collaborative work.

DOI: 10.4324/9781003397977-16

Exploring Culturally Responsive Science Teaching in the CRSE PLG

We came together out of a shared passion and understanding that an important need was not being met in our New York City science classrooms. We formed the Culturally Responsive and Sustaining Education Professional Learning Group (CRSE PLG) as we started our third year of teaching, because we wanted to figure out how to teach our students better. We were curious about CRSE, which was something we had heard about—but we hadn't yet gained a clear understanding of how it might play a role in our science classrooms. We were hopeful that learning about CRSE with colleagues might help us to more effectively teach our racially, culturally, and linguistically diverse students.

Through our time together in the group we have come to see CRSE as essential for supporting students' connections to science. Figuring out how to find the right answer on an exam can be important. That is the world we live in. But we can also help students draw science into their personal and communal webs of knowledge by getting to know them. In doing so, we need to understand our students' perspectives, and view what they know and what they enjoy as assets for our teaching and their learning (López, 2017; Moll et al., 1992). This does not replace the high expectations that we hold for our students' learning (Ladson-Billings, 1995). Rather, it allows us to challenge our students and encourage their learning in more personal and culturally responsive ways.

In this chapter, we reflect individually as well as communally on our learning through our participation in the CRSE PLG. At the end of this chapter, we share "A Gift for Our Readers," a resource we have developed over our years together to support our efforts in implementing culturally responsive and sustaining curriculum and instruction in our science class-rooms. We hope this tangible product will help in expanding your CRSE-based pedagogy.

Beginning to Explore CRSE

In the early days of our work together, even while the phrase *culturally relevant education* was buzzing around education circles, we found little information on culturally relevant *science* education. Most of the research was based on connecting to students by including perspectives challenging dominant narratives, and was most prevalent in other disciplines such as language arts, the humanities, and social studies. In addition, over time, the terminology evolved to include the concepts of *responsive and sustaining* (see Chapter 1). We see all of these terms—relevant, responsive, and sustaining—as naming

Figure 11.1 An early CRSE PLG meeting at the American Museum of Natural History in September, 2018. Photo by Elaine Howes.

perspectives that help us move toward the inclusion of all students in our science classrooms. But still, in our explorations of this growing field, *science teaching* was rarely forefronted. (See Figure 11.1.) We strongly believe that CRSE should not be limited to language arts, humanities, and social studies classrooms, and changing this has been a prime focus in our work together.

To us, the insanity of the lack of attention to CRSE in science teaching is that science isn't just random facts that are unrelated to humans and their lives. Rather, science is rooted in the context of our imperfect social and cultural worlds. Because of this, for a lot of the science content that we teach, there are real-world connections that we can delve into and consider through a humanizing lens (see Chapters 1, 6, and 10). Especially when there is a strong social aspect to the curriculum, students can clarify and value their own perspectives while learning to appreciate the lived experiences and viewpoints of people whose backgrounds are very different from their own.

We have also come to the realization that CRSE requires that we understand not just our science content, but also our audience. Who are our students? What do they bring to our science classes, and to science as a human endeavor? How do their lived experiences influence what and how they understand science? How can we use what we learn about our students to help engage them in science learning?

CRSE Helps Students Learn the Practices of Science

We believe science is a great platform for exploring CRSE in all its many aspects. For example, it can provide a starting place for teaching about observations and questioning, which are foundational to science. We can work with our students to build on the idea that inferences and hypotheses are not just educated guesses but are built on observation skills and background knowledge, and even our preexisting biases. For instance, "reading the room," a social skill that many students have already honed, may allow them to understand how and why science is done. *What do you see when you walk into a room? What are you looking for? What do you recognize? What do you wonder about?*

We also believe that CRSE can support students in learning to ask scientific questions without the fear of being "wrong." Students' questions about their world can be encouraged rather than shut down, leading to richer conversations and incorporation of additional science concepts. Science is about observing the world around us and wondering why things are the way that they are. Helping students to recognize that their own expertise and their questions about the world are the grounding for scientific thinking has been an important part of our inquiry into CRSE.

CRSE Is Student-Centered

Likewise, we have come to believe that CRSE in science includes supporting students to think, discuss, observe, critique, and work towards improving the world around them. Although the various members of our professional learning group have different perspectives and thoughts on what CRSE looks like in our science classrooms, one agreed-upon idea is that CRSE must be student-centered. Students, especially those who have been historically alienated from the school structures and systems in our country, bring voices, experiences, and perspectives that we aim to elevate and interests we hope to cultivate (Lyiscott, 2019). We, as teachers, must provide ways for students to share who they are and what they want. Along these lines, we can have students plan investigations, design experiments, and decide the direction of a curriculum while holding them to high academic standards in ways that connect to their identities and draw on their assets.

We have, in our lifetimes, seen shifts in who participates in Western science. Furthermore, we know that diversity is foundational to finding solutions to the many challenges facing our world. We hope to help our students see science in this way, and we believe that CRSE is helping us to do that.

Individual Reflections on Our Work Together

At this point in this reflection chapter, we decided that each of us would write about why we decided to commit to this group and what it means for us and our teaching. As we all bring different perspectives, we believe that individual reflections can provide insight without filtering them through a single lens. We have organized our thoughts with these questions in mind:

- Why did I join the CRSE PLG?
- How did the CRSE PLG support my thinking and learning?
- What do I wish I had known about CRSE as a novice teacher?
- What is it like to conduct teacher research?

Why Did I Join the CRSE PLG?

Kin: When I first joined the CRSE PLG, I was teaching in my second school. This school's student population was primarily emerging bilingual and multilingual learners, with two years or fewer in the USA. I felt grossly unqualified, that nothing that I did would ever be good enough, and that I was letting students down. I joined this group with the hope that I could figure out some ways to better support them. Having this collaborative group, where we all struggled to try to be better teachers, was invaluable in my growth as a new-ish teacher. I don't think that I would have been able to continue teaching if we had not had this time and support, both mentally and emotionally. The classroom environment and learning are never straightforward. What works one year may not work the next and, for that, I appreciated having people with whom I could bounce ideas off of. I appreciated having people with the same goals in mind.

Maya: I have mostly been able to float through my life unaware of my whiteness. I was living the definition of white privilege. But when I was in graduate school in Puerto Rico, I woke up. Attending a science conference, I noticed the demographics represented white people, like me, and not people of Color, like my graduate school colleagues. I finally saw that science needed a face lift.

When I was invited to join the CRSE PLG, I knew that the answer I was looking for was within reach. Curriculum that put my students at the center, gave them agency over their own learning, involved them in real science experiences, and taught them how to use science as a tool for justice … What more could I want? That isn't to say that everything after that moment was a walk in the park. For one thing, I found myself spending hours looking for images and videos that both featured people of Color and effectively explained the concepts I was trying to teach. (I should add that this was not because people of Color made ineffective videos, there just were very few in any of the science videos.) I had to rewrite entire units to reframe the narrative around real societal issues that evoked a critical stance. The hardest part was that there were very few resources available to make this process easier. That's why it felt so important to be part of the team working to change that.

How Did the CRSE PLG Support My Thinking and Learning?

Kin: I especially valued the practice of storytelling that we called Stories from the Field (or "Stories"). That structure gave us a time to tell our own stories and reflect on our CRSE teaching practice. Constant reflection gave me the opportunity to build my own teaching ethos, which I will simplify as "care about the students." Even during the many times that I did not have much to say during Stories, listening to others go through their own struggles and successes gave me opportunities to consider what I might do next time. There is a lot of struggle with teaching that we do not openly talk about. We do what we can for the students in front of us, but it never feels like enough. My blanket goal of better science teaching was a way for me to try to bridge that gap.

Sam: I joined the CRSE PLG long distance from Los Angeles, right at the beginning of the quarantine in April/May 2020. As in other urban centers, inequality among students was heightened with the sudden switch to virtual learning. As we scrambled to provide students with computers and meals, I felt helpless trying to continue our regular lessons through Zoom calls. This group became a lifeline for me, providing not only a space to vent but also ideas on how to ensure our students were safe and supported. Participation in the CRSE PLG has changed my approach to teaching forever. Since writing my chapter, I have already expanded my approach by tying our curriculum to the city in which our school is located and where my students live. For example, instead of teaching a water cycle lesson based on a fictional town, I've focused on New York City's water supply and sewage system. This has opened up new opportunities for partnerships with community organizations that I may not have discovered without this group.

Sean: This time together in the group has been incredibly important for hearing how my close friends and colleagues have approached CRSE in their classrooms. I always get ideas of things I want to try. For example,

Figure 11.2 CRSE PLG members celebrating after leading a workshop on CRSE in science teaching at the Noyce Summit in Washington D.C., July, 2019. Photo by Elaine V. Howes.

my sensory openers about weather variables were inspired in part by things Arthur has spoken about in these meetings. As others have mentioned, it's also great to bounce ideas off each other. I feel like these conversations have helped me better understand my goals as a teacher and how to make my lessons more student-centered. (See Figure 11.2.)

What Do I Wish I Had Known About CRSE as a Novice Teacher?

Sean: If I were a new teacher starting out, what I would want to know about CRSE and science teaching is to put a priority on getting to know your students and building relationships with them. This helps inform your teaching and is the important groundwork for teaching any content. Design lessons early in the year and, in each unit, draw on students' assets and their natural curiosity.

Collaboration and reflection are so important for teachers at any stage of their career. I appreciate immensely having my colleagues' support in things that I now see as central to CRSE.

Rooting our science lessons in students' assets and communities, engaging students in the processes and practices of doing science, and possibly the most challenging for a science teacher, realizing, acknowledging, and addressing the concept that science is not impartial and that it exists within a cultural context.

Raghida: To me, CRSE is the thread that connects school to so-called "real life." It is meeting every student where they are, and helping them feel seen and heard in the content and classroom culture. CRSE in the science classroom includes student-designed experiments, student-written procedures, student-led discussions about content and history and race and socio-economic status. Most of all, CRSE in the science classroom includes anecdotes and explanations that connect science to everyday life without tokenizing.

It is most helpful to know that CRSE is not a checklist. There are some days where I am proud of a classroom that amplifies student voice, provides multiple modalities for accessing and assessing content, and gives students choice. There are other days where only one strong CRSE practice is present and that's totally okay. I tell myself that everyone is still learning. This reminder helps me keep my patience with my students, as I see myself as a learner, just like them.

What Is It Like to Conduct Teacher Research?

Susan: In my previous scientific work I was an engineer; I knew hydrology very well, and I knew from where to draw resources to back up what I was seeing and I knew I could use historical data to explain the current situation more clearly. For teacher research, coming up with a research question—and then actually considering, "How do I research that?"— felt the most difficult to me. It wasn't like an experiment I could repeat several times and validate the results. This would not be possible in the classroom, where I adjust what I'm teaching based on the students I have at that time, so even when I teach the same lesson twice, it's different … .

But doing teacher research reinforced for me that one of the most important things about teaching is the aspect of reflecting. I've never known any other profession where reflection is a part of it. It seems that in many jobs, you're going to do a design (e.g., an engineering design), you're going to review it, you're going to make sure your calculations are right, you're going to hand it to somebody else, they're going to check your calculations, and then it's either going to get approved or sent back for revision. But with teaching it doesn't work that way. It's really incumbent on me, the teacher, to reflect deeply and honestly about what is working and what's not working for my students. I felt that my research and writing was the deepest form of reflection on teaching that I had ever done.

Kin: I see myself as a teacher, not a researcher. For me, teacher research was never the goal. It was an avenue for me to try to do better and all of it was hard. The hardest part though, the part that I struggled with most, was the idea of thinking of my students as subjects of research. There were times when I felt like I was a failure for using my classes as a way to further my own understanding of education. While I do not consider myself as a researcher, I tried to do what I could for the students in front of me.

To that end, there were two things that I found most interesting and helpful in terms of research, and they worked well in conjunction with each other. The first was the literature review that we started in our group's infancy. We would take time in our meetings to read different articles and sometimes we had articles to read on our own time that we would then debrief with the rest of the group. Learning new things, discussing them, and trying to figure out ways to best implement them felt like something I could do that was actually helpful.

Maya: I know that we've talked about this a lot—how hard it was to think of myself doing research when it wasn't traditional scientific quantitative research. So I was really struggling. It didn't really feel like doing research. It felt like just trying to be a good teacher, which also goes back to the idea of what is CRSE or how do we do CRSE and how is this important. But that also made it a little bit easier—the fact that it was research made it less emotionally challenging for me to go through the days where it didn't really go so well. It allowed me to really reset my frame of mind. Because I was telling myself, "You know, Maya, this is an experiment. You had an idea. And now you are finding out if that idea is something that is going to work or not. And the only way you're going to learn is if you test it. And we know how it went. And it wasn't always great. But when you do make it quantitative, and you put it on a graph, you see that it actually did work out." So it was a good learning opportunity. In that sense, it was a really good way for me to get to know my kids, and get to know my kids' needs and how I could step back and make it less teacher-centered and more student-centered, and really hear their voices.

Arthur: My teacher research and learning about CRSE has not been an independent endeavor. My entire experience is with a learning community. I think that was critical. I don't like to bail on things. But I probably wouldn't have decided to engage in something as rigorous if I didn't have a team to hold me accountable. I think it would have made me set the bar a little bit lower.

So if you want a collaborative endeavor then seek out a team to challenge you. I don't know how you can do that on your own, and I don't know how you can gather a team, but if you can find a way—start spreading out the Captain Planet rings.[1]

Collaborative Reflection: There Is No Lockstep for Individual Teacher Change

Each of us changed in different ways because it depended on what we cared about and focused on. For some that was curriculum development, for others it was relationship building, or finding avenues for students to show who they are, or schoolwide development, or giving students a time to interpret real life and how it connects to science. Some of us have recently left classroom teaching, taking on different ventures and moving to other states. Some of us wonder if we could do more good elsewhere. But we can all agree on the urgent need to help our students gain confidence in their voices and the assets they bring to the science classroom by learning about and using these in our science teaching. We also hope to help them see that science can be used to support the kinds of social change that they hope to make in their world.

For us, learning about CRSE required the support of the group. Our work went from focusing on little aspects of what we can do in the classroom (things like small activities to engage student voice) towards something more akin to a framework of what we want our classrooms and our curriculum to look like. Additionally, in exploring the extensive literature on CRSE, having a group allowed for a range of literature exploration that would have been unthinkable to tackle alone. None of this would have been possible without our ongoing collaboration.

Coming from science backgrounds, we found qualitative research, let alone studying one's own classroom, to be quite daunting. We have also concluded that teacher research is unavoidably intertwined with teaching itself.

A Gift for Our Readers: A Resource for Working on CRSE in Science Classrooms

To provide materials and ideas for the kind of teaching that encourages students to bring their own understanding of the social and natural worlds into the science classroom, we looked for stories and scenarios. We searched first in the existing literature about CRSE, and then among colleagues in our own networks (e.g., schools, science education organizations, listservs). We then created a spreadsheet of literature and classroom activities that could mesh with our understanding of CRSE science teaching. (See Figure 11.3.) We have shared this resource with preservice teachers, inservice teachers, and teacher educators locally and nationally in professional learning workshops and at conferences. This is a living document, and we continue to make updates as we encounter new resources. (For a current version of this resource, see Appendix.)

To illustrate, we present a few examples using a question and response format and highlight particular CRSE tenets that shine through. The samples that we share provide a sense of the ways that science can be taught in more culturally responsive and sustaining ways. Please understand that this is a part of a living document, and that these examples will change as the world does, and as our students do.

Promotion of Social Awareness

Students are provided with opportunities to connect learning to social, political, or environmental concerns

Guidance is provided by connecting students to members of the community who are working in STEM fields and who represent their cultural identities.
- Visiting places like the ASRC at CUNY and Columbia Engineering

Exposure to diverse communities who are affected by issues with science inquiry opportunities (chemistry, geology, medicine, natural resources).
Examples:
- Flint water crisis - there are numerous lesson plans such as this one by the Global Oneness Project - Do We Have a Right to Clean Water?
- Mari Copeny - youth activist for water rights - Asked and Answered: President Obama Responds to an Eight-Year-Old Girl from Flint
- Pipeline through Native American lands - The Youth Activists Behind the Standing Rock Resistance (with Lesson Plan)
- Impoverished communities at risk from global warming (unable to move out of flood zones: exposure to increased health and disease issues)
  - Climate Refugees
- Lead poisoning in NYC: NYCHA
- Testing lead water in schools with NYC DEP

Non-dominant populations and their strengths and assets are included, so that students of diverse race, class, gender, ability, and sexual orientation can relate and participate fully.
Examples:
- People in Puerto Rico recovering from hurricanes despite very limited help from the US government
Strategies and lesson plans - advance the human rights of all people.
- https://www.learningforjustice.org/

Figure 11.3 A portion of our *Components of Culturally Responsive and Sustaining Education in the STEM Classroom* resources guide, based on the *Culturally Responsive Curriculum Scorecard* from NYU Metro Center/Steinhardt (2019). Screenshot by Elaine V. Howes, a resource developed by the CRSE PLG.

Question 1: How can I teach about water resources in a more culturally responsive and sustaining way?

CRSE Tenets Highlighted: Value what students bring to the classroom as assets and use them as resources for teaching and learning. Adopt and support students in developing a critical stance toward sociopolitical structures and processes.

A CRSE Response

Open up the conversation: What do students bring to the study? Culturally responsive and sustaining teaching about the Flint, Michigan, water crisis requires a deep understanding and appreciation of students' backgrounds, experiences, and strengths. The process starts with building background knowledge, assessing what students already comprehend about water resources, and blending these understandings into the Flint crisis. Start with having students share their own experiences and observations related to water quality and access. Ask them what they know about water resources, why they are important, or if they are familiar with any local water issues. Maybe some students have experienced their own water crisis with a boil water advisory. Some may have had friends or family who lived in Flint during the crisis, or can share stories from their backgrounds about water scarcity or pollution. Provide prompts for students to share their beliefs or practices about water preservation and cleanliness. This is a good way to potentially tap into their cultural backgrounds.

Explore resources: **What can we learn about the sociopolitical elements of this crisis from those affected most?** Look for documentaries, news articles, and interviews about the Flint crisis that feature a variety of diverse voices from the people directly affected in the community. This will help you to frame the crisis within the context of social justice and equity and promote empathy by discussing how socioeconomic status, race, and other factors affected the response to the crisis. This is also a way to support students in developing a critical stance through encouraging deep dives into sociopolitical processes and structures through multiple perspectives.

Use collaborative learning strategies: **How can I engage my students in studying the crisis and developing possible solutions and/or actions?** Use collaborative learning and group projects so that students can draw on their diverse strengths. For example, have them create a water conservation campaign, develop a contamination crisis response plan, or research potential risks inherent to their own local water system. Challenge them to think critically about the handling of the crisis, the possibilities for different outcomes if they were in charge of the response, and have them plan prevention strategies that empower young people to engage in social and environmental change.

Question 2: How can I leverage my students' assets and honor their lived experiences when I teach about earthquakes?

CRSE Tenets Highlighted: Draw upon students' cultures to strengthen and sustain their connections to them. Adopt and support students in developing a critical stance toward sociopolitical structures and processes.

A CRSE Response

Recognize that students bring experiences and knowledge. Consider, how I can welcome students' experiences and knowledge in a respectful way? Students from seismically active regions can offer unique experiences and perspectives, thereby enriching classroom discussions. Provide a safe space for students who may have (or have had family who) experienced a traumatic earthquake, allowing them to share their stories. Encourage these students to share their communities' measures for earthquake preparedness or other safety strategies. If none of the students have experienced an earthquake, consider showing a video to help them envision what it could be like.

Use your study of earthquakes to introduce cultural connections by discussing how earthquakes are handled in different parts of the world, thereby emphasizing the global aspect of seismic phenomena. Earth science (and other sciences) are about natural phenomena that are global in nature, experienced by humans all over the world. Teaching about earthquakes, for example, can provide a context for understanding how diverse communities perceive, prepare for, and respond to natural disasters, thus forging a stronger connection between classroom learning and real-world scenarios. Relevant earthquake preparedness strategies that directly impact students' lives and resonate with their cultures could include discussions on local building codes, community emergency planning, or traditional safety measures for earthquakes. To further reinforce real-world and community connections, you may choose to invite into your classroom local public officials or community members who have experienced earthquakes as guest speakers.

Question 3: How can I support my students in developing a critical stance when talking about climate change?

CRSE Tenets Highlighted: Adopt and support students in developing a critical stance toward sociopolitical structures and processes. Draw upon students' cultures to strengthen and sustain their connections to them.

A CRSE Response

Support students in exploring how climate change does not affect everyone equally. Draw on students' prior knowledge by asking if any of them have observed or experienced any effects of climate change. *Do they know anyone who has lived through an unseasonably intense storm, or faced flooding due to rising sea levels? Have any of them noticed the heat island effect in their city, or the increased droughts in their rural area?* You can leverage these personal experiences to connect to more global case studies. For example, the island nations such as Kiribati, Maldives, and Tuvalu will likely be completely underwater by the year 2100. Even though the people from those countries have contributed the least to the climate crisis, they are suffering the most. Use maps and other resources to look at the island nations that are already experiencing the adverse effects of rising sea levels. Alternatively, explore the effects of climate change on your local setting, with guiding questions such as: *Which neighborhoods are going to be most affected? Why? What are the possible effects?* You and your students can look at the health statistics of various communities in addition to maps to see what is already happening in terms of health (e.g., childhood asthma) and safety (e.g., urban flooding, extreme heat).

Explore and develop strategies for climate resilience. Have students read about the creative, resourceful strategies people are adopting to protect themselves and combat climate change. Students could read firsthand accounts from climate refugees-turned-advocates to practice empathy for their experiences and see themselves as scientists and activists, while enhancing their literacy skills.

Another way students can learn about the effects of climate change is through sharing their own lived experiences. *What similarities exist between urban and rural environments? What is different? Why? How have systemic inequities led to these differences? How have communities come together to overcome these challenges?* Students can then practice a critical stance by working in groups to design a solution for climate change in their own community. If this activity is framed through the lens of politics and the economy, they can also develop practical skills of self-advocacy and budgeting.

Question 4: How can I hold high expectations for all students' academic learning when teaching about volcanoes?

CRSE Tenets Highlighted: Hold high expectations for all students' academic learning. Draw upon students' cultures to strengthen and sustain their connections to them.

A CRSE Response

Maintain high expectations for all students. Pose questions to find out what your students already know about volcanoes to learn what you can expand upon. Include voices of students' who have a cultural connection or a lived experience with volcanoes or volcanic eruptions. Include resources from students' everyday lives, such as from video games they play (e.g., mining obsidian in *Minecraft*) or popular books from their grade level. Have students research and present a case to their peers on a volcanic eruption and the social and economic impacts of volcanic eruptions on the surrounding areas. Make sure that students include different points of view in their presentations. Have students research the school community to see if a volcanic eruption can possibly occur in their location, and then debate the answer using evidence. If the school is in danger, challenge students to an engineering problem by creating a volcanic eruption warning system and evacuation route or plan for the appropriate city. If the school is not in a volcanic eruption zone, have students write a scientific explanation to the public about why they should not fear volcanoes.

Note

1 Captain Planet rings, from the television series *Captain Planet and the Planeteers,* are sent to young people all over the world so that they can participate in saving the Earth from environmental destruction. See Wikipedia: https://en.wikipedia.org/wiki/Captain_Planet_and_the_Planeteers

References

Ladson-Billings, G. (1995). But that's just good teaching! The case for culturally relevant pedagogy. *Theory Into Practice, 34*(3), 159–165.

López, F. A. (2017). Altering the trajectory of the self-fulfilling prophecy. *Journal of Teacher Education, 68*(2), 193–212.

Lyiscott, J. (2019). *Black appetite. White food.: Issues of race, voice, and justice within and beyond the classroom.* Routledge.

Moll, L. C., Amanti, C., Neff, D., & Gonzalez, N. (1992). Funds of knowledge for teaching: Using a qualitative approach to connect homes and classrooms. *Theory Into Practice, 31*(2), 132–141.

12

The Ties That Bind

Jamie Wallace, Elaine V. Howes, and the CRSE PLG

Introduction

This book features individual and collaborative teacher inquiries into culturally responsive and sustaining education (CRSE) in our science classrooms. While each study takes place in a science classroom in New York City, each setting and environment is distinct.

There are two purposes intended for this chapter. First, we review what we see tying the chapters together. We have learned from working on CRSE in our science classrooms through this long-term collaboration, and we explore several intersecting threads woven through our inquiries. We ground these threads both in their connections to CRSE and existing research. Second, drawing on our experiences, we provide some insights and advice for teachers contemplating or embarking on teacher research into CRSE. We hope that you may find this useful in your journey. We intentionally do not use the word final, as our thoughts on CRSE can never be final as we perpetually continue learning …

DOI: 10.4324/9781003397977-17

Exploring the Threads That Weave Throughout

In this book, we structured our individual inquiries (Chapters 2–9) into three categories: *Exploring Instructional Strategies for CRSE, CRSE and the Science Classroom*, and *School-Based Teacher Collaboration Reflecting CRSE*. These are accompanied by three *Collaborative Chapters: Stories and Reflections*—an analysis of our reflections on our stories at the beginning of the pandemic (Chapter 10) and two reflection chapters (Chapter 11 and 12). While the authors' have different interests, range of perspectives, and varied explorations, several intersecting threads run through the chapters. In these threads, we see ideas that were developed through our ongoing group discussions. Each author is responsible for their own chapters, and they also exist within a web of ideas and experiences shared by members of the CRSE PLG. It is the threads that make up this web that we attempt to delineate in this chapter.

Another way to describe this collaborative writing venture is as a patchwork quilt, where all members of the group provide patches that are stitched together to make a creative collaborative composition. Moreover, like a patchwork quilt containing unique and individual efforts culminating in a larger endeavor to provide warmth and care, our chapters come together to serve a larger purpose—to examine, document, reflect on, and share our collaborative journey in this work. For us, the threads that tie together our quilt across teacher inquiries include: *learning with and from students: the importance of relationships; trust; reflecting on the teacher learning journey;* and *identity, positionality, and self-reflection.* Like the seams of a quilt, we recognize the overlaps and nuances across these threads which serve to reinforce and strengthen the ties that bind them together. We briefly look at how these strands appear throughout the larger fabric.

Learning with and from Students: The Importance of Relationships

A strong thread running throughout the chapters is the vital importance of relationships with students and the notion of learning with and from our students. The importance of student–teacher relationships is not a unique idea and has been studied for decades (Bernstein-Yamashiro & Noam, 2013; Brinkworth et al., 2018; Knight-Manuel & Marciano, 2018; Thornberg, 2022; Wubbels & Brekelmans, 2005). However, through the inquiries, stories, and reflections that we share in this book, relationships are centered by the teachers and viewed through the lens of culturally responsive science teaching.

Forefronting teacher–student relationships in *Humanizing Science Teaching through Building Relationships* (Chapter 6), Kin describes how knowing his students (and their knowing him) is not only central to his teaching but also an important feature in developing his classroom environment. His teaching journey through various high-need public high schools, each with a distinct school culture, inspired him to prioritize making his classroom "a welcoming and open space for students to learn and to be themselves." He describes choosing to focus his energies in his classroom and his relationships with students, because that is where he can make a difference. This perspective views students as individuals with unique and complex lives (aspects of which he shares in his chapter). Through the lens of CRSE, Kin uses an approach that insists that he see his students as humans first, recognizing the whole person and the importance of their well-being, and noting that "there can be no true checklist for CRSE because the work itself means that you work with the students and who they are." In reflecting on his practice, he describes a type of humanizing pedagogy (del Carmen Salazar, 2013; Kazembe, 2022; Mehta & Aguilera, 2020). He shares how, for him, it was more important to learn about the individual person than about their cultural group.

Learning with and from her students, Susan (Chapter 7) highlights multiple strategies that she uses as starting points to develop deeper individual connections. She describes the role of relationships in her teaching and in CRSE as foundational, considering it as potentially "the most impactful for both students and teachers." It is through the individual relationships that she forms with her students that she learns more about what is important to them and their goals for the future. In doing so, she values these aspects as assets and uses them to inform her science teaching (one of the CRSE tenets), such as integrating minerals in makeup into lessons particularly to resonate with her students who are interested in becoming aestheticians. In a similar manner, Caity (Chapter 8) recognizes the importance not only in learning about who her students are in the moment but in who they want to become, an asset-based view that is critical to CRSE.

Looking at relationships through a different perspective, Sam (Chapter 3) describes her learning about her sixth-grade students, comparing challenges to form connections during remote teaching days to the return to in-person teaching. During the many months of remote teaching, there were times when we would not see students' faces, let alone learn about their favorite kind of backpack, what they liked to draw, and how they interacted with one another when together in a physical classroom. Sam notes that going back in person after months of remote teaching helped her to realize how important the day-to-day interactions with her students were, especially in learning about their interests and the assets that they bring to the classroom. Similarly, the isolation and loss of human connection was keenly felt across the teachers in their reflections on the stories they told during the pandemic (Chapter 10). Through analyzing our pandemic teaching stories, we also learned that during this period of upheaval and trauma, centering students' well-being was paramount and surpassed any other priority.

Trust

The importance of establishing trust in the classroom is a strand that weaves through several chapters. Research supports this notion, confirming teachers' beliefs that building a culture of trust is necessary in order to create an inclusive classroom (Samuels, 2018). While a key feature in enacting CRSE, there is potential at times for it to be overlooked or taken for granted. Across each chapter, we see various approaches and strategies used to set up a learning environment where trust can be cultivated—trust between teachers and students, students and their peers, teachers and their peers, and with school leadership and administration.

Explaining her teaching philosophy (Chapter 8), Caity remarks that establishing trust is essential in order for students to learn in her science classroom and for her to learn about her students. She describes routines and activities that she uses in her classroom to build trust before delving into science content, such as lesson openers, restorative circle practices, and identity mapping. These approaches and strategies are used to deliberately bring healing-centered work into her science classroom in efforts to build trust and create safe spaces necessary for students to feel welcome, confident, and further develop their identities as scholars.

Trust also connects deeply with Kin's inquiry, *Humanizing Science Teaching through Building Relationships* (Chapter 6). Kin explains how he grounds his teaching practice in connection to CRSE in honesty and building trust, which is necessary for developing and maintaining relationships. He contends, "If there is a loss of trust then there becomes a loss of respect and that teacher-student relationship becomes fractured." Building off this duo of honesty and building trust, but using different terms, Susan (Chapter 7) describes how she needs to be genuinely herself with her students "to build mutual respect" in order for a relationship to become "a two-way street." This aspect of being genuinely oneself with students also surfaces across multiple chapters as critical to establishing trust.

From another angle, Maya (Chapter 2) writes to her fictional student, Dante, explaining that she understands why he and his peers may be distrusting of science and science education. Drawing on historical racial inequities in science, Maya explores what it means to cultivate a "scientist identity," especially for students from underrepresented groups. In addition to acknowledging that there may be a lack of trust in science classrooms, we see this as an exercise in trust as Maya shares with a student the imperfections of the scientific enterprise, which can be a risk for a science teacher. She aims, through this choice, to help a student who expresses indifference toward science to see that he is already thinking in scientific ways and that it is possible to push against prevailing stereotypes and consider becoming a professional scientist. In this, Maya asks Dante to trust her as she helps facilitate learning, both about Earth science and developing a critical stance. Similarly, several other chapters also recognize that trust has to be present in the learning environment in order for anyone to feel comfortable in taking risks. For instance, as Raghida shares (in Chapter 5), "The classroom learning environment has to include trust and the normalization of mistakes in order for students to feel capable of being vulnerable and taking intellectual chances."

As Hammond (2014) points out, trust is at the center of positive relationships and caring is the way in which we generate trust to build relationships. Similarly, trust is necessary to feel safe and can thus be viewed as a component of establishing a welcoming and affirming environment, one of the principles in the NYS Education Department's CR-SE Framework (2019).

Reflecting on the Teacher Learning Journey

Woven underneath the surface across chapters is the strand of the teacher journey and the learning that happens along the way. Integral to teacher research and to teaching in general, reflection is a central practice for this journey and what is learned from it (Cochran-Smith & Lytle, 1993; Schön, 1987). Indeed, critical reflection is also necessary for understanding and implementing culturally responsive and sustaining pedagogy (Durden & Truscott, 2013; Howard, 2003). In this way, we consider reflection an essential practice that ties together CRSE and teacher research. Thus, it is not surprising that "reflection" as a research method (Kraft, 2002) comes across in each inquiry. While each author does this in distinct ways in their studies, it functions as a unifying seam in our quilt, holding the individual patches in place.

Reflecting on her journey into her own learning as a teacher, Susan (Chapter 7) speaks directly about her struggle in figuring out what CRSE *means* and how she applies it in her own teaching. Her journey in this regard is a reckoning that results in a decision to ground her work with CRSE in 3 Rs, a construct that she uses as a framework to make her teaching moves in this area more concrete. Through her efforts to understand her students better so she can teach them science in ways that engage them and they find relevant, Susan makes a tectonic shift from focusing on students' motivations in school to honestly inquiring into what it is that interests them in life, and what their goals are for themselves after graduation. Across her reflections, the notion of co-creation of experiences and learning together with her students shines through as she describes approaches through reciprocal storytelling and sharing accounts of her own lived experiences as well as hearing about those of her students.

With the goal of understanding shifts made in pedagogical approaches that benefited students during the chaos of the COVID-19 pandemic in order to to carry them forward, Arthur, too, engages in ongoing reflection on practice as a method to investigate his inquiry (Chapter 9). Arthur shares his own reflection on his teaching together with those of his teacher team, stepping back to look at instructional approaches from a schoolwide lens, interwoven with his beliefs and philosophies about learning and education.

In many ways, Kin's inquiry (Chapter 6) is a deep dive into his learning journey as a teacher as he reflects on his experiences and practice in different schools. In doing so, he strives to understand what has been working for his students, and asking what features of his classroom environment are within his control. He also considers through this journey the teaching "moves" he makes to cultivate a learning environment that is inviting. It was through reflection, supported by survey data and stories, that Kin identified salient features of the classroom environment that he consistently seeks to establish, grounded in conversation, honesty and trust building, and responsibility and opportunity.

Identity, Positionality, and Self-Reflection

Threaded throughout the chapters is the underlying notion that it is important to examine our own identities when engaging in CRSE. This is part of our continuing journey in which we consider how our identities frame and shape how we perceive and navigate the world. Exploring one's own identity is also a necessary and critical aspect of engaging in research. This is especially evident in connection to CRSE, where we need to be self-reflective and examine our own identities (Gay & Kirkland, 2003; Gorski & Dalton, 2020), challenge our interpretations and assumptions, critique our biases (Fenge et al., 2019; Milner, 2007), and consider the dynamics of conducting research within and across one's culture (Merriam et al., 2001). Through these chapters, we see an acknowledgment that research, like teaching, is not neutral; that we all have lenses through which we make sense of the world; that concepts of race, culture, power, and privilege are embedded in and have implications for the work that we do.

This is apparent in *Building Relationships to Support Relevance* (Chapter 7), as Susan engages in critical self-reflection, where she challenges herself and raises questions about how being a white woman might perpetuate inequities and what it means to approach CRSE from a place of privilege. In her learning journey, she noticed how "blinders started to fall off," recognizing her inherent biases and thinking critically about societal structures.

Exploring how he seeks to foster his students' connections with the natural world, Sean (Chapter 5) recognizes that his love for science stems from his white, suburban, middle-class upbringing. He acknowledges that his passion for science grew through his family road trips into nature and learning in informal science settings. In a subtle manner, he considers his privilege. He thinks about how his experiences and upbringing are very different from his students' lived experiences.

In *Crisis Precipitates Change* (Chapter 9), Arthur considers the dynamics of power and privilege that come into play in the classroom and how they might affect the interactions and relationships he has with his students. At the crux of his chapter, Arthur questions the notion of authority that the teacher in the classroom holds and seeks to promote change in hopes of shifting the power dynamics to further center students.

In thinking about trauma-informed and healing-centered approaches (Chapter 8), Caity shares a story relating an activity that made her stop and consider how she needed to "dig deep" into her own identity. In reflecting on her identity, she thinks about the privileges that she was afforded as a white person and the role that race can play in constructs such as opportunities, work, sense of belonging, and safety. She describes how she approaches identity work with her students, both in her science classroom and in her crew, and considers what it means to be comfortable to share about identity markers with others. She questions what it might mean to engage in this work as a white teacher whose students are primarily young people of Color, many who identify as LGBTQIA+. Additionally, she expresses the concern that she is "not confident that I am doing everything right. In fact, I believe I am doing a lot of things wrong."

In considering the parts of her identity that she brings to the classroom, Raghida (Chapter 5) shares anecdotes with her students that are grounded in her experience as a child of immigrants and a multilingual learner. Bringing aspects of her identity into her classroom, she describes an exercise to support her students in thinking critically about race and identity in science, drawing on her own experience in her undergraduate program. In perusing photos of predominantly white male professors on the program website and encouraging her students to make observations, she uses this activity to tap into assumptions and biases related to identity in science.

Across chapters, many of the teachers articulate that their own identities do not mirror the majority of those of their students racially, ethnically, and, in some cases, socioeconomically. In considering her ethnic heritage, Sam (Chapter 3) has an insight about her deep cultural connection to the neighborhood where her school with a predominantly Latine student population is located as she learned that it is the same place where her grandparents lived when they emigrated from Ireland. This propels her to consider how neighborhoods change over time in terms of racial/ethnic composition and with gentrification, as she also reflects on the demographic shifts at her old high school. Similarly, Sean (Chapter 4) addresses the dissonance between his upbringing as a white male who grew up in a suburban setting from the experiences of his students who are Latinx, living in an urban environment, and many of whom are children of immigrants.

In the introduction and collaborative chapters, we briefly discuss how the CRSE PLG group identifies. We recognize that many of the members of the group identify as white or white presenting. This has been a focus of innumerable discussions and debates throughout the years (see photo in Chapter 1 of us reading and discussing *White Fragility* during the first year of the CRSE PLG). Indeed, we even had conversations about whether we should publish this book considering our racial composition. Likewise, as facilitators, Elaine and Jamie have engaged in additional conversations not only addressing their identities as white women and teacher educators but also interrogating how power plays a role in the group's dynamics (Wallace et al., 2022). As a group, we strive for transparency, honesty, and humility in our work and our roles. We also see the vital importance of centering the expertise of teachers and the need to publicly share teacher research into CRSE, particularly in science.

Engaging in this work can be challenging and uncomfortable—and it should be. This is the way to move forward and to work toward social change and transformation. (See Figures 12.1 and 12.2 for photographs of the CRSE PLG considering possible threads and ties across inquiries.)

Figure 12.1 CRSE PLG members making a representation of the threads that go throughout their chapters. Photo by Elaine V. Howes.

Figure 12.2 CRSE PLG members reviewing a representation they had made of the threads that go throughout their chapters. Photo by Elaine V. Howes.

Advice for Teachers Interested in Teacher Research on CRSE

Understanding that this work is challenging and takes time, we wanted to reflect on our own experiences working with the CRSE PLG and offer some insights to teachers who might be interested in doing something similar. The following are ideas and kernels of advice that we hope you might find useful as you consider diving into teacher research or inquiries into CRSE.

What advice would you offer teachers interested in embarking on teacher research?

Starting teacher research on CRSE

- My advice is to not think of it as research. Do it as a way to learn about how you can best help the students. - **Kin**
- Remember that your own journey in learning is an important part of your research and that support is there, whether in the form of a

professional learning group, validation in the literature, in informal posts and discussions, or even just in casual conversation. Finally, remember that your experiences are valuable and can be a helpful part of another's education journey. - **Kin**

- [CRSE] is not a quick fix. I think that might be one of the hardest things to get past, that we really want to figure it out quickly. And we really want to find the answer. You know we've been doing these research projects for years and we still haven't found a simple answer. - **Maya**
- Some advice to others starting out with CRSE and with research is "trust the process." To be patient with it and with yourself. - **Maya**
- Do research when you care about something, when you have data that's telling you something needs to be done. - **Arthur**

Methods

- It was difficult to figure out an area to focus my research that would be worthwhile to investigate. You have to have a passion about something. Something will inevitably come up that you have opinions or questions about or an interest in. - **Arthur**
- Keep a journal of these things that happen, and over time a pattern emerges. As a science teacher, you're always looking for patterns. If you've got some notes and you're looking at them, it kind of talks to you. - **Arthur**
- Log everything and try to learn from everything. - **Kin**
- I wouldn't have been able to keep track of any of this or really to see the journey if I didn't take notes throughout the process. - **Maya**
- It was hard at times to go to conferences or present what we were doing, but at the same time, the feedback that we got was another way for us to have outside eyes asking us questions, and that was also in its own way a form of reflection. - **Susan**

Form or find a group

- Even if your school has incorporated specific actions around CRSE, see if other teachers and staff are interested in joining you in creating a professional learning group. You can start with reading a book like the short guide *Transforming Our Public Schools: A Guidebook to Culturally Responsive Education* by NYU's Metro Center, and progress to a full book study. (See other readings that we have found useful in Chapter 1.) - **Susan**

- Find a good group that you can work with. Find some mentors or facilitators. There's so much literature out there right now that I feel that it's really important, but it can be time consuming to really dig into. And that's why again having a group is important. I like how there were times when our group would have each person pick one chapter to be responsible for, and bring that back to the group. Like when we did the Ta Na-hesi Coates (2015)—that was one that really struck me as important, and an ah-ha moment for me. And then I could go more deeply into that and then bring that back to the group.
  - Susan

References

Bernstein-Yamashiro, B., & Noam, G. G. (2013). Teacher-student relationships: A growing field of study. *New Directions for Youth Development, 2013*(137), 15–26. 10.1002/yd.20045

Brinkworth, M. E., McIntyre, J., Juraschek, A. D., & Gehlbach, H. (2018). Teacher-student relationships: The positives and negatives of assessing both perspectives. *Journal of Applied Developmental Psychology, 55*, 24–38. 10.1016/j.appdev.2017.09.002

Coates, T.-N. (2015). *Between the world and me*. Spiegel & Grau.

Cochran-Smith, M., & Lytle, S. L. (Eds.). (1993). *Inside/outside: Teacher research and knowledge*. Teachers College Press.

del Carmen Salazar, M. (2013). A humanizing pedagogy: Reinventing the principles and practice of education as a journey toward liberation. *Review of Research in Education, 37*(1), 121–148. 10.3102/0091732X12464032

Durden, T. R., & Truscott, D. M. (2013). Critical reflectivity and the development of new culturally relevant teachers. *Multicultural Perspectives, 15*(2), 73–80. 10.1080/15210960. 2013.781349

Fenge, L. A., Oakley, L., Taylor, B., & Beer, S. (2019). The impact of sensitive research on the researcher: Preparedness and positionality. *International Journal of Qualitative Methods, 18*, 1–8. 10.1177/1609406919893161

Gay, G., & Kirkland, K. (2003). Developing cultural critical consciousness and self-reflection in preservice teacher education. *Theory Into Practice, 42*(3), 181–187. 10.1207/s15430421 tip4203_3

Gorski, P. C., & Dalton, K. (2020). Striving for critical reflection in multicultural and social justice teacher education: Introducing a typology of reflection approaches. *Journal of Teacher Education, 71*(3), 357–368. 10.1177/0022487119883545

Hammond, Z. (2014). *Culturally responsive teaching and the brain: Promoting authentic engagement and rigor among culturally and linguistically diverse students*. Corwin Press.

Howard, T. C. (2003). Culturally relevant pedagogy: Ingredients for critical teacher reflection. *Theory Into Practice, 42*(3), 195–202. 10.1207/s15430421tip4203_5

Kazembe, L. D. (2022). A mighty love: Culture, community, and liberatory practices mong educators of color. In C. D. Gist & T. J. Bristol (Eds.), *Handbook of research on teachers of color and Indigenous teachers* (pp. 713–725). American Educational Research Association.

Knight-Manuel, M. G., & Marciano, J. E. (2018). *Classroom cultures: Equitable schooling for racially diverse youth*. Teachers College Press.

Kraft, N. P. (2002). Teacher research as a way to engage in critical reflection: A case study. *Reflective Practice, 3*(2), 175–189. 10.1080/14623940220142325

Mehta, R., & Aguilera, E. (2020). A critical approach to humanizing pedagogies in online teaching and learning. *International Journal of Information and Learning Technology, 37*(3), 109–120. 10.1108/ijilt-10-2019-0099

Merriam, S. B., Johnson-Bailey, J., Lee, M. Y., Kee, Y., Ntseane, G., & Muhame, M. (2001). Power and positionality: Negotiating insider/outsider status within and across cultures. *International Journal of Lifelong Education, 20*, 405–416. 10.1080/02601370120490

Milner, H. R. (2007). Race, culture, and researcher positionality: Working through dangers seen, unseen, and unforeseen. *Educational Researcher, 36*(7), 388–400. 10.3102/0013189X07309471

NYC Culturally Responsive Education Working Group and the Education Justice Research and Organizing Collaborative (EJ-ROC) at the NYU Metro Center (n.d.). Transforming our public schools: A guide to culturally responsive-sustaining education. https://static1.squarespace.com/static/5bc5da7c3560c36b7dab1922/t/5ed12955d45eb54e7a0854a3/1590765951611/CEJ_CRSEBook_v7.pdf

New York State Education Department (NYSED). (2019). *Culturally responsive-sustaining education framework*. http://www/nysed.gov/crs/framework

Samuels, A. J. (2018). Exploring culturally responsive pedagogy: Teachers' perspectives on fostering equitable and inclusive classrooms. *SRATE Journal, 27*(1), 22–30.

Schön, D. A. (1987). *Educating the reflective practitioner: Toward a new design for teaching and learning in the professions*. Jossey-Bass.

Thornberg, R., Forsberg, C., Hammar Chiriac, E., & Bjereld, Y. (2022). Teacher–student relationship quality and student engagement: A sequential explanatory mixed-methods study. *Research Papers in Education, 37*(6), 840–859. 10.1080/02671522.2020.1864772

Wallace, J., Howes, E., Funk, A., Krepski, S., Pincus, M., Sylvester, S., … & Swift, S. (2022). Stories that teachers tell: Exploring culturally responsive science teaching. *Education Sciences, 12*(6), 401.

Wubbels, T., & Brekelmans, M. (2005). Two decades of research on teacher–student relationships in class. *International Journal of Educational Research, 43*(1–2), 6–24. 10.1016/j.ijer.2006.03.003

Appendix
Components of Culturally Responsive and Sustaining Education in the STEM Classroom

This document has been developed collectively, first by the American Museum of Natural History's Culturally Responsive and Sustaining Education Professional Learning Group (CRSE PLG), then through contributions from teachers at workshops and meetings. While this list is by no means comprehensive, it serves as a launching point for educators seeking resources to bring culturally responsive and sustaining education into their STEM classroom. It was based on the 2019 Culturally Responsive Curriculum Scorecard (Bryan-Gooden et al., 2019). When we created this document, a scorecard was only available for English Language Arts classes, so we modified the criteria to be relevant for STEM classes. In the time since, NYU has released a scorecard for STEAM.

Please note that all links were checked on August 4, 2023. Given that the Internet is constantly in a state of change, we cannot guarantee the longevity of these resources and websites.

Promotion of Social Awareness
Students are provided with opportunities to connect learning to social, political, or environmental concerns.
Guidance is provided by connecting students to members of the community who are working in STEM fields and who represent their cultural identities. • Visiting places like the Advanced Science Research Center (ASRC) at City University of New York (CUNY) and Columbia Engineering
Exposure to diverse communities who are affected by issues with science inquiry opportunities (chemistry, geology, medicine, natural resources).

Examples:

- Flint water crisis - there are numerous lesson plans such as this one by the Global Oneness Project - Do We Have a Right to Clean Water? (https://www.globalonenessproject.org/lessons/do-we-have-right-clean-water)
- Mari Copeny - youth activist for water rights - Asked and Answered: President Obama Responds to an Eight-Year-Old Girl from Flint (https://obamawhitehouse.archives.gov/blog/2016/04/27/asked-and-answered-president-obama-responds-eight-year-old-girl-flint)
- Pipeline through Native American lands - The Youth Activists Behind the Standing Rock Resistance (with Lesson Plan) (https://www.kqed.org/lowdown/27023/the-youth-of-standing-rock)
- Impoverished communities at risk from global warming (unable to move out of flood zones; exposure to increased health and disease issues) (https://www.epa.gov/climateimpacts/climate-change-and-health-socially-vulnerable-people)

  o Climate refugees

- Lead poisoning in NYC: New York City Housing Authority (NYCHA)
- Testing lead water in schools with NYC Department of Environmental Protection (DEP)

Acknowledgment of scientists and innovators of diverse backgrounds. Examples: (as you add to this collection consider adding a heading such as women in the sciences; youth as innovators and activists; etc.)

Women in the Sciences
- Breakthrough: Portraits of Women In Science - Explore our short-film anthology that follows women working at the forefront of their fields. (https://www.sciencefriday.com/spotlights/breakthrough/)

  - Breakthrough: The Killer Snail Chemist - the work of Mandë Holford, an associate professor of chemistry and biochemistry at The Graduate Center of The City University of New York (GC/CUNY) and Hunter College is featured by Science Friday. (https://www.sciencefriday.com/videos/breakthrough-killer-snail-chemist/)
  - Protecting the Waterways of the Navajo Nation and Breakthrough: Bitter Water - The work of University of Arizona researcher Karletta Chief, assistant professor and assistant specialist in the University of Arizona Department of Soil, Water, and Environmental Sciences, is featured as part of a screening of short

> films by Science Friday. (https://www.sciencefriday.com/segments/protecting-the-waterways-of-the-navajo-nation/; https://news.arizona.edu/calendar/110641-film-breakthrough-bitter-water)

- Meet 12 Badass Scientists … Who Also Happen to Be Women https://fellowsblog.ted.com/meet-12-badass-scientists-who-also-happen-to-be-women-ace8d797bcad
- United States Geological Survey geophysicist Dr. Rufus Catchings. https://www.usgs.gov/news/one-first-black-usgs-geophysicists-pioneers-subsurface-research
- *NY Times* has a series of obituaries that got missed when the women died - a lot of these are female scientists. (https://www.nytimes.com/interactive/2019/obituaries/black-history-month-overlooked.html)

Youth as Innovators and Activists

- https://youthactivismproject.org/
- https://www.insider.com/young-activists-climate-change-guns-greta-thunberg-2019-9
- Women in Science: 50 Fearless Pioneers Who Changed the World, written and illustrated by Rachel Ignotofsky (Publisher: Ten Speed Press; Illustrated edition (July 26, 2016))
- Women of Impact (National Geographic) (https://www.nationalgeographic.com/culture/article/how-women-have-changed-the-world-covered-in-special-issue)

Environmental and health problems faced by people of Color or females are presented.
Examples:

- Increasing number of heat waves in NYC disproportionately affects people of Color
- Is there a statistically higher risk of dying for mothers and babies of Color during childbirth? If so, why? - *NY Times* article "Maternal Mortality Rate in U.S. Rises, Defying Global Trend, Study Finds" (https://www.nytimes.com/2016/09/22/health/maternal-mortality.html)
- Are mothers of Color more likely to die during childbirth in the United States than white mothers? (https://www.ncbi.nlm.nih.gov/pmc/articles/PMC1595019/)
- *Period. End of Sentence.* (2018 documentary short film) The documentary short follows a group of local women in Hapur, India, as they learn how to operate a machine that makes low-cost, biodegradable sanitary pads, which they sell to other women at

affordable prices. This not only helps to improve feminine hygiene by providing access to basic products but also supports and empowers the women to shed the taboos in India surrounding menstruation – all while contributing to the economic future of their community. (https://www.netflix.com/title/81074663)
- NYCHA lead paint (https://www.nytimes.com/2018/11/18/ny-region/nycha-lead-paint.html)

Non-dominant populations and their strengths and assets are included, so that students of diverse race, class, gender, ability, and sexual orientation can relate and participate fully.
Examples:

- People in Puerto Rico recovering from hurricanes despite very limited help from the U.S. government

Strategies and lesson plans - advance the human rights of all people.
- https://www.learningforjustice.org/

Asset-Based Instruction
Educators value students and what they bring to their classrooms.

Diverse student identities are seen as assets and strengths that can advance individual and group learning, rather than seen as challenges or difficulties to be overcome.

- Group Learning Protocols: Designed to address all learning needs and convey value to every student voice in the room through collaboration and accountable talk
  - o Examples: Rumors, Domino Share, Elbow Exchange, Show & Share, etc. (https://www.alled.org/group-learning/gather-responses/rumors/)

The curriculum communicates an asset-based perspective by representing people of diverse races, classes, genders, abilities, and sexual orientations through their strengths, talents, and knowledge rather than their perceived flaws or deficiencies.

- Hazard preparation project. Have students create a safety pamphlet/public service announcement/Infographic based on hazards that they have already experienced, so they can share how they prepared for them.

o   Optional: ELLs can translate their safety pamphlets into their home language for extra credit (https://cdn.kqed.org/wp-content/uploads/sites/26/2017/10/When-disaster-strikes-lesson-plan.pdf; https://www.readwritethink.org/classroom-resources/lesson-plans/picture-worth-thousand-words-0)

The curriculum recognizes the validity and integrity of scientific advances based in communities of color, collectivist cultures, matriarchal societies, and non-Christian religions.

- Teach about contributions to science by cultures that are non-European. (https://www.pewresearch.org/science/2022/04/07/black-americans-views-of-and-engagement-with-science/; https://www.verywellmind.com/what-are-collectivistic-cultures-2794962)

Guidance is provided on opportunities to engage students' families and/or community assets to enhance lessons, with an understanding that students' home and family lives are diverse.
Examples:

- Finding people within the *community* to represent science to students (e.g., Emergency Medical Technicians (EMTs), restaurant owners, city planners and community leaders)
- What science skills do they use in everyday lives (and parsing science skills vs. science content)

The diversity of the scientific community is reflected in the curriculum and it is made known that science is open to anyone who is curious about the natural world and who wants to take a scientific approach to his/her/their investigations.
Examples:

- ALL students are referred to as scientists.
- Examples of women's contributions to science that have been overlooked.
  - o   Women star analysts at Harvard (https://www.theatlantic.com/science/archive/2016/12/the-women-computers-who-measured-the-stars/509231/)
  - o   *Hidden Figures* curriculum (https://journeysinfilm.org/product/hidden-figures/)

- Examples of non-European contributions to science

  o Historical representations of celestial objects (Lone Dog Calendar "Year the Stars Fell," Mawangdui Silk Texts, Moroccan Petroglyphs, History of Celestial Objects at European Southern Observatory) (https://www.eso.org/public/events/astro-evt/hale-bopp/comet-history-1/; https://americanindian.si.edu/sites/1/files/pdf/education/poster_lone_dog_final.pdf; https://www.newworlden-cyclopedia.org/entry/Huangdi_Sijing; https://www.researchgate.net/publication/330401781_Rock_Art_and_Arc-haeoastronomy_in_Morocco_Preliminary_Observations)

- Students are presented with opportunities to use science to solve their own problems Interesting site by UNESCO - Dive into intangible cultural heritage!

  o https://ich.unesco.org/en/dive

Social situations and problems are not seen as individual problems but are situated within a societal context.
Examples:

- Research into Seneca village (African American community) in Central Park. (https://projects.mcah.columbia.edu/seneca_village/)
- Flint water/Dakota Pipeline (https://www.globalonenessproject.org/lessons/do-we-have-right-clean-water; https://www.kqed.org/lowdown/27023/the-youth-of-standing-rock)
- Which communities are at most risk of flooding due to sea level rise?
- Exploring health issues within their community in the societal contexts

  o Air pollutant levels and asthma rates (https://www.ncbi.nlm.nih.gov/pmc/articles/PMC4465283/)

- How science skills such as data collection and conclusion writing can be used to aid in combating social problems

Science is inherently a social endeavor that depends upon the collaboration of highly diverse people from a variety of backgrounds.
Examples:

- Most scientists work in labs or field stations, surrounded by other scientists and students.
- Scientists often collaborate on studies with one another, mentor less experienced scientists.

Students come to the classroom with knowledge and skills in areas of their own interests. Some of these interests and skills include references to pop culture they enjoy outside of the classroom setting.

- *Anime Series - Cells at Work!* Episodes and description Human Biology (https://thescienceof.org/cells-at-work/)
- *Anime Series* - Moyashimon Benefits of microbiome (https://en.wikipedia.org/wiki/Moyasimon:_Tales_of_Agriculture)
- *Video Game* - Topographic Maps in The Legend of Zelda Video Game (ex: http://mtwcanimation.com/hyrule)
- *Video Game* - Minecraft Educational Game - Rocks and Minerals (https://education.minecraft.net/en-us/lessons/museum-math-rocks-and-minerals2)
- *Anime/Game* - Life Imitates Pokemon: The Virtues and Necessities of Technology-Based Peer Education in Today's Schools (https://eric.ed.gov/?id=EJ600581)
- *Comic Book Builder - Pixton* - Online comic builder for students (https://www.pixton.com/welcome)
- *Video Game* - Roblox Education Guide (https://d5m0vnpvja3z4.cloudfront.net/assets/blt93d4c160ce74d3da/InspirationGuide-RobloxEducationPartnersJune21.pdf)

Curriculum Design
Avenues are defined within the curriculum that allow students to access their culture.

The curriculum does not communicate negativity or hostility toward people of marginalized backgrounds through verbal or nonverbal insults, slights, or snubs.

- Avoiding Microaggressions in the Classroom (https://www.slu.edu/cttl/resources/resource-guides/microaggressions.pdf)

The curriculum presents different points of view on the same event or experience, especially points of view from marginalized people/communities.

- Debating the Future Use of the Arctic: Stakeholders & Role Playing — Have students represent different stakeholders in the Arctic (Indigenous peoples, oil companies, shipping companies, animal rights groups, environmental activists, etc.) and debate from their perspectives how the Arctic should be used, especially in the context of melting sea ice and consequential new shipping routes (https://www.cfr.org/teaching-notes/teaching-notes-emerging-arctic)

The curriculum encourages students to take actions that combat inequity and/or promote equity within the school or local community.

- Research projects that have an authentic audience of community members (i.e., pamphlets to be distributed, a health fair)
  o Useful websites for creating deliverables: Canva.com, Flipgrid, Adobe Spark (https://info.flip.com/en-us.html)

Curriculum and instructional activities promote or provoke critical questions about the societal status quo. They present alternative points of view as equally worth considering.
Examples:

- Researching scientists of underrepresented demographics (e.g., for a bulletin board or project) then facilitating a discussion about why it is necessary to do this kind of research and why it can be challenging to conduct

Curriculum includes references to different ethnic and cultural traditions, languages, religions, names, and clothing. This may include the diversity of Ancient Culture's responses to scientific challenges and their unique cultural assets.
Examples:

- Ancient Chinese seismographs (https://www.engadget.com/2018-09-28-backlog-zhang-heng-seismoscope.html)
- Native American Astronomy (https://pubs.aip.org/physicstoday/article-abstract/37/6/24/434752/Native-American-astronomyArc-haeoastronomers-are?redirectedFrom=fulltext)

How cultures across the world have seen their myths and legends in the stars - which is a pretty cool site as well.

- http://www.datasketch.es/may/code/nadieh/
Science Friday - Ways of Knowing - Written in the Stars
- https://www.sciencefriday.com/articles/indigenous-peoples-astronomy/
- http://annettelee.com/index.php/media/
- http://annettelee.com/index.php/portfolio/star-maps/

Instructional Supports for Teachers:
Guidance is provided on how teachers can reflect the cultures of the student population.

Provide teachers with tons of strategies on how to do that including examples from videos.

Guidance is provided on customizing and supplementing instruction to reflect the cultures, traditions, backgrounds, and interests of the student population.

Guidance is provided on being aware of one's biases and the gaps between one's own culture and students' cultures.

- Have students or teachers complete true/false activities, or questionnaires that address their biases before a new topic or unit.
- Access various books, articles, interviews, podcasts from those of different perspectives and identities to expose oneself to new learning about other cultures.

Guidance is provided on making real-life connections between academic content and students' cultures, local neighborhood, environment, and resources.

- Showing students through activities in the classroom that science affects their lives every day, even when they don't think about it or when they think that science isn't relatable to them
- Using readings from recent news or local current events to frame a particular topic or provide more context
  o Useful websites: Newsela.com, Science News For Students
  o NovelNY (www.novelnewyork.org)

Guidance is provided on giving students opportunities to contribute their prior knowledge and experience with a topic, not just respond to the text and information presented in class.
Examples:

- Asking students to share personal experiences with any given topic
- Do case studies of students' experiences (e.g., choose someone's hometown for investigative phenomenon)
- Eliciting students' ideas to explain phenomenon rather than explaining immediately

- Predict, Observe, Explain protocol (https://arbs.nzcer.org.nz/predict-observe-explain-poe)

Guidance is provided on engaging students in culturally sensitive experiential learning activities.
Examples:

- Restorative Justice Activities
- Trauma Informed Instruction

Guidance includes expectations of diverse responses, as long as they are supported by scientifically valid evidence and reasoning.
Examples:

- Allow students' interpretation of terms: for example "Big Bang" and "revolution" may have strong meaning to students and their constructs may not "fit" within the astronomy definitions but may allow for valid idea exchanges.
- Should people be allowed to construct new buildings in seismically active regions?

Developed based on the model of the *Culturally Responsive Curriculum Scorecard* from NYU Metro Center/Steinhardt (Bryan-Gooden et al., 2019), with additions from *A Framework for K-12 Science Education: Practices, Crosscutting Concepts, and Core Ideas* (NRC, 2012).

References

Bryan-Gooden, J., Hester, M., & Peoples, L. Q. (2019). Culturally responsive curriculum scorecard. New York: Metropolitan Center for Research on Equity and the Transformation of Schools, New York University.

National Research Council (NRC). (2012). A framework for K–12 science education: practices, crosscutting concepts, and core ideas. National Academies Press.

Name Index

Note: *Italic* page numbers refer to figures. Page numbers followed by "n" indicate material in numbered endnotes

Adams, A. D. 19
Aderin-Pocock, M. 54
Adjapong, E. S. 56
Aguilera, E. 129, 216, 253
Alderman, D. 213
Alim, S. xiii, 6, 15, 17, 18, 84, 113, 154, 212
Alliance for Excellent Education 145
Almanac 47
Almeida, C. A. 145
American Indian Alaska Native Tourism Association 47
Amplify Education Inc. 79n1
Anderson, R. 54
Arbery, Ahmaud *219*, 225, 226
Armento, B. J. 45
Asiksoy, G. 195

Bai, H. 223, 228, 229
Baines, J. 18, 45
Bal, A. 220
Baldwin, J. 25
Bali, M. 191
Ball, A. F. 13
Bang, M. 15, 19, 21
Baroutsis, A. 50
Barron, H. A. 16
Barakat, R. 59
Barnard, A. 12
Bateson, M. C. 213
Bell, M. 24
Bettez, S. C. 21
Beauchamp, C. 213
Bernstein-Yamashiro, B. 252
Berryman, M. 220
Bintz, J. 22
Blancas, E. 55
Bloom, B. S. 144

Blum, G. I. 216
Borko, H. 22
Boutte, G. 21
Bozkurt, A. 216
Brayboy, B. M. J. 15
Brekelmans, M. 252
Brinkworth, M. E. 252
Bronx Zoo 80n2
Brown, B. A. 17, 21
Brown, C. T. 16
Brown, D. F. 69
Brown, M. R. 45
Brown-Jeffy, S. 16
Bryan-Gooden, J. 15, 263, 272
Buckley-Marudas, M. F. 216

Calabrese Barton, A. 17, 19, 21
Calvert, S. 137
Campion, N. 47
Carpenter, T. P. 52
Carpinetti, A. 55
Carter Andrews, D. J. 15, 216, 230
Center for Youth Wellness 172
Chand, R. 216
Chatterjee, H. 86
Christianakis, M. 220
Chen, G. 135
Chowning, J. 190
City of New York, 154, 215
Clandinin, D. J. 213, 214, 228
Coates, T.-N. 25, 261
Cochran-Smith, M. 6, 12, 22, 23, 24, 27, 255
Conklin, H. G. 13
Connelly, F. M. 213
Cooper, J. E. 16
Cooper, M. 215

Topic Index

Note: *Italic* page numbers refer to figures. Page numbers followed by "n" indicate material in numbered endnotes

For Product Safety Concerns and Information please contact our EU
representative GPSR@taylorandfrancis.com
Taylor & Francis Verlag GmbH, Kaufingerstraße 24, 80331 München, Germany